EMERGENCY MANAGEMENT

THE WILEY BICENTENNIAL–KNOWLEDGE FOR GENERATIONS

Each generation has its unique needs and aspirations. When Charles Wiley first opened his small printing shop in lower Manhattan in 1807, it was a generation of boundless potential searching for an identity. And we were there, helping to define a new American literary tradition. Over half a century later, in the midst of the Second Industrial Revolution, it was a generation focused on building the future. Once again, we were there, supplying the critical scientific, technical, and engineering knowledge that helped frame the world. Throughout the 20th Century, and into the new millennium, nations began to reach out beyond their own borders and a new international community was born. Wiley was there, expanding its operations around the world to enable a global exchange of ideas, opinions, and know-how.

For 200 years, Wiley has been an integral part of each generation's journey, enabling the flow of information and understanding necessary to meet their needs and fulfill their aspirations. Today, bold new technologies are changing the way we live and learn. Wiley will be there, providing you the must-have knowledge you need to imagine new worlds, new possibilities, and new opportunities.

Generations come and go, but you can always count on Wiley to provide you the knowledge you need, when and where you need it!

WILLIAM J. PESCE
PRESIDENT AND CHIEF EXECUTIVE OFFICER

PETER BOOTH WILEY
CHAIRMAN OF THE BOARD

EMERGENCY MANAGEMENT

Concepts and Strategies for Effective Programs

LUCIEN G. CANTON

WILEY-INTERSCIENCE
A JOHN WILEY & SONS, INC., PUBLICATION

Published by John Wiley & Sons, Inc., Hoboken, New Jersey.
Published simultaneously in Canada.

For general information on our other products and services or for technical support, please contact our Customer Care Department within the United States at (800) 762-2974, outside the United States at (317) 572-3993 or fax (317) 572-4002.

Wiley also publishes its books in a variety of electronic formats. Some content that appears in print may not be available in electronic format. For information about Wiley products, visit our web site at www.wiley.com.

Library of Congress Cataloging-in-Publication Data is available.

Canton, Lucien G.
Emergency Management: Concepts and Strategies for Effective Programs

ISBN-13: 978-0-471-73487-1
ISBN-10: 0-471-73487-X

Printed in the United States of America.

10 9 8 7 6 5 4 3 2 1

For

David Fowler
1944–2001

Newsman, colleague, mentor and friend,
But, always, a true San Franciscan

CONTENTS

PREFACE

An invasion of armies can be resisted, but not an idea whose time has come.
—Victor Hugo

This is not the book I planned to write. When I first started this project in 2004, my original plan was to base the book on material I had developed for a course on emergency planning and management that I had taught for the University of California at Long Beach and to incorporate new information I had gained by teaching seminars on the National Preparedness Standard for New York University's INTERCEP program. It was to have been a very conventional book, focused on the tactical issues so dear to us old dinosaurs in emergency management.

Hurricane Katrina changed all that. As my colleagues and I wrestled with the issues of catastrophic response, I began to question a lot of our traditional approaches to disaster response. An article by Dr. E.L. Quarantelli on the qualitative differences between catastrophe and disaster added fuel to the flames and sent me back to the social science literature. A comment during a presentation at the 2005 International Association of Emergency Managers conference provided the catalyst that made all the disparate elements come together.

This book approaches emergency management from a different perspective than the traditional four phases of emergency management. It does not have the equally traditional listing of hazards and impacts. It doesn't even give a lot of detail about response. There are plenty of excellent books already available that deal with these topics. Instead, I have focused on the development of emergency management programs and attempted to position these programs within local government in a way that contributes to community goals by helping to manage community risk. The idea of emergency management as an enterprise-wide program forms, I believe, the core of the National Preparedness Standard, NFPA 1600, and its derivative, the Emergency Management Accreditation Program (EMAP) Standard. This concept also demands a change in the role of the emergency manager, from that of a technical expert who is responsible for everything vaguely related to disasters, to that of a program manager who coordinates the community's management of risk.

This is a very different perspective from the way we have traditionally viewed emergency managers. However, the best emergency managers have either already adapted to this concept or are on their way to doing so. More importantly, an entirely new generation of future emergency managers is emerging from our educational institutions, potential leaders who are trained in this new paradigm. The real issue for our profession will be gaining the acceptance of elected officials and the public for this new role and overcoming the roadblocks created by the well-meaning but out-of-touch Department of Homeland Security.

The ideas in this book are likely to be controversial and, I hope, spark discussion among my colleagues. There is no single best way to respond to disasters—by virtue of the need for innovation and creative problem solving during response, there really can't be. However, we can define a common set of criteria that positions us for success. It is this belief that has caused a number of my colleagues to expend considerable efforts to develop NFPA 1600 and the EMAP Standard. So while some of my ideas may be controversial, they are grounded in this common set of criteria and in a considerable body of social science research.

I had hoped initially to write a book that would be applicable to both the public and private sectors. However, I began to realize that there are, in fact, qualitative differences between the two sectors that make such a task extremely difficult. The principles are the same but there are enough subtle nuances that would have made the book cumbersome. Wherever I could, I have tried to focus on concepts and principles, so it is my hope that this book may be of some value to my colleagues in the private sector.

I've had to make similar decisions in some of the titles I selected. Over the last few years there have been many new players getting involved in disaster response. We are seeing a convergence of disciplines that will have a profound impact on our professions in the future. Risk managers, security managers, business continuity managers and so many others are lending important new skills to our programs. So while I have focused this book on public sector emergency managers, it is my hope that there will be applicability to the other disciplines that are involved with disaster response. Each of these disciplines has a specialized body of knowledge that makes us experts in our field but there is a commonality among disciplines when we start discussing emergency preparedness and response.

It's been said that no book is the sole product of the author and how true that is! The two most humbling things I know are teaching a course or writing a book about your profession. It forces you to confront how little you really know and how much better others have expressed the ideas you're groping toward. Sir Isaac Newton once said, "If I have seen farther than others, it is because I have stood on the shoulders of giants." Many of those giants are listed in the bibliography.

I am an emergency management dinosaur. I came to the profession with previous experience in private security and the military at a time when you learned your craft from your mentors and your colleagues. It was years before I discovered that everything that I had learned the hard way had already been written about by social scientists like Russell Dynes and E.L. Quarantelli. The advantage to coming to their work late in my career is that I know they are right—their work corresponds to the lessons I have learned in over 30 years of dealing with crisis. So to all those social scientists that are building the knowledge base so critical to our profession, at least one emergency manager has heard you and appreciates your hard work.

For the rest, there are friends and colleagues around the world who have taught me my craft and had a part, however unknowing, in the writing of this book. I am always amazed at the generosity of my fellow emergency managers and their willingness to help in any and all circumstances. If I have learned anything in this business, it has been because of you. You know who you are.

A final thank you goes, as always, to my wife Doreen, who suffered through multiple rereadings of the initial manuscript and encouraged me to keep plugging away. For over 20 years she has been my moral compass and best friend.

INTRODUCTION

"I cannot imagine any condition which would cause a ship to founder. I cannot conceive of any vital disaster happening to this vessel. Modern ship building has gone beyond that."
—Captain Edward John Smith, Commander of Titanic

When on Friday, February 24, 2006, the White House issued a report entitled, The Federal Response to Hurricane Katrina: Lessons Learned, one salty emergency manager observed, "It ain't a lesson learned until you correct it and prove it works. Until then, it's just an observation." The White House report is just one of several released around the same time, all saying essentially the same thing: as a nation, the United States is not prepared to deal with catastrophe.

How is this possible? The United States has been in the emergency management business for over 50 years. There are volumes of social science reports on human behavior in disaster. There are detailed records on historical disasters that have occurred in the past 300 years and geological records going back to prehistory. Millions have been spent on building the capacity to respond. Since September 11 there has been an even bigger push to strengthen and enhance emergency response capabilities. And yet, in the biggest test in U.S. history, the system failed at all levels of government.

Hurricane Katrina is not, unfortunately, an isolated case. Every disaster seems to generate a list of failures couched as "lessons learned," along with pledges to improve the system. Few of those pledges are ever implemented. Yet, like Captain Smith, citizens in the United States believe that a sophisticated system of response is in place to protect them from the unthinkable. There is an expectation that, no matter what the event, government will be there to provide immediate and effective relief.

To a certain extent, emergency management in the United States is a victim of its own success. Response is extremely fast compared to other countries and there is a culture of professionalism among first responders that makes them second to none. However, this has led to the expectation that disaster response is a government responsibility, not a collective one, and there are increased demands for more immediate and detailed services. This is a demand that has obvious limits, as demonstrated by Hurricane Katrina.

Government officials at all levels go out of their way to reinforce these public expectations. In a recent speech to the Heritage Foundation, Michael Chertoff,

Secretary of the Department of Homeland Security, encouraged people to be prepared, saying, "…you cannot count on help coming in the first 24 or even 48 hours of a catastrophe…people who are prepared with that kind of planning do much better if they have to wait 24 to 48 hours than people who don't do that planning." Secretary Chertoff seemed unaware that he had just shortened the normal recommendation of preparing for a minimum of 72 hours by 24 to 48 hours and further encouraged the public's expectation of immediate response.

It is unheard of for an elected official to admit the truth. Disasters, by definition, overwhelm available local resources. You can never be fully prepared—there isn't enough money or political will to fund all the requirements for mitigation and preparedness planning that would ensure full preparedess. Preparedness is a balancing act, with must politicians betting that a disaster will not happen on their watch and that the public will not discover the thin veneer that passes for preparedness. Jurisdictions are unprepared and it is extremely unlikely that they will ever reach the level of preparedness that the public believes already exists. The bar has been set too high to be supported by local, state, or federal government without a major shift of priorities.

However, as one reads after-action reports and "lessons learned" one begins to sense commonalities. It is seldom the critical initial life-saving response that is criticized. Police, fire and emergency medical personnel usually get high marks for their efforts in a crisis. Witness the praise deservedly heaped on the U.S. Coast Guard for its rescue of 33,000 victims during Hurricane Katrina (an operation so successful that Secretary Chertoff now believes that only Coast Guard admirals are qualified to serve as Principal Federal Officials). Instead, criticism seems to fall into two areas. The first is related to traditional victim services such as sheltering or evacuation. Criticism of victim services usually reflects inadequacy of service or confusion in the delivery of services brought on by poor coordination among relief agencies. Indeed, some social scientists suggest that the biggest problem in disasters is not the impact of the event on the victims but the lack of coordination among multiple responding agencies. This confusion and lack of coordination can actually impede the delivery of services.

The second major area of criticism relates to long-term issues. This is usually characterized by conflicts over reconstruction policies. Again, one notes concerns over confusion in the process. There is a lack of coordination and public participation that leads to delays in the rebuilding of a community and the restoration of its economic base. It is during this recovery period that one generally sees the emergence of finger pointing and an increase in underlying social tension. There is usually a conflict between citizens who want to rebuild quickly and return the community to the way it was and officials who push for improved structures or social re-engineering.

Again, one has to ask the question, "Why?" Why, in a system that has 50 years of experience in countless disasters, that has national guidelines, that has millions in government funding, that has reams of textbooks and social science reports, why is it that the system seems to fail more than it works and why do those failures always seem to be in the same areas? Can the U.S. do better?

The fact that these failures seem to occur in almost every disaster and in almost always the same areas would seem to suggest that there is something wrong with the system. Social science suggests some of the reasons. Emergency management issues do not generally engage local officials. In most jurisdictions, the responsibility for developing emergency response capacity rests with a single individual and is usually an additional duty. Emergency planning is viewed as a task centered on the development of a paper plan and there is no real linkage between emergency management and community goals and vision. Worse, emergency plans incorporate assumptions based on disaster myths and do not reflect the reality of human behavior in disasters.

This book suggests that the United States can do better by changing the nature of emergency management and traditional response. It is time for a different approach, one that is supported by social science and by new national standards for emergency management programs. This approach is based on the concept that emergency management is a distributed process, one that must be collectively performed by the community. This suggests that emergency management must be integrated with other community goals and as such must be perceived as adding value to the community. This added value is achieved by helping the community manage overall risk.

The community-wide approach also holds implications for the emergency manager. Instead of being a technical expert on emergency operations, the emergency manager becomes a program coordinator whose job is to facilitate the development of a community strategy for managing risk and to oversee the enterprise wide implementation of that strategy. This focus on strategy allows all the various components of the community to work together to achieve a common vision of resilience.

The first three chapters of this book focus on the three pillars on which successful emergency management is based: an understanding of history, knowledge of social science research, and technical expertise in emergency management operations. The chapters also provide insight as to how emergency management has evolved and suggests reasons why the current method of response planning doesn't work as well as it should.

Chapter 4 discusses establishing and administering the emergency management program. Traditionally, emergency management "programs" have merely been a collection of activities with only vague relation to each other, primarily driven by federal grants. Chapter 4 provides a mechanism for addressing program governance and oversight and for linking program elements through a strategic plan.

Chapter 5 considers the analysis of risk as the basis for strategy development. It considers both the traditional macro view of hazard identification and analysis as well as the micro view required for continuity planning.

Chapter 6 covers strategy development, a major weakness in many emergency management programs. The focus is not so much on individual strategies as it is on the interface between the strategies. It is this conceptual basis that helps build the flexibility needed in disaster response.

Chapters 7 and 8 focus on the development of the various plans needed within the emergency management program. Again, the chapters are more concerned with the process than with the specific plans and with tactical planning issues rather than field operations. Chapter 7 discusses planning concepts while Chapter 8 suggests methodologies to translate these concepts into actual plans.

Chapter 9 considers issues related to tactical response. It discusses the pros and cons of incident management systems and suggests a coordination methodology that may prove more effective than traditional command and control structures.

Chapter 10 focuses on the roles and responsibilities of senior officials in the management of strategic response. It suggests that the normal involvement of the senior officials in the emergency operations center may be counter-productive and suggests new ways of managing disasters using crisis management principles.

Together, the chapters make a case for a change in how emergency management programs are integrated into communities and in the role of the emergency manager. These changes are consistent with the direction of the National Preparedness Standard and the current guidance provided by the Department of Homeland Security and are supported by social science research. It is hoped that they might point toward a more effective system of disaster response.

Chapter 1

EMERGENCY MANAGEMENT: A HISTORICAL PERSPECTIVE

What experience and history teaches us is that people and governments have never learned anything from history, or acted on principles deduced from it.
—Georg Wilhelm Friedrich Hegel

Since the dawn of time, humankind has had the need to deal with crisis of all types. For much of history, this response was personal and intimate and the victim did not always survive the encounter. As people grouped together for the common good, the idea of some sort of collective response to crisis gradually evolved. As governments came into being, this idea became an expectation that it was part of the responsibility of government to provide protection and assistance in times of crisis. From this expectation came the discipline of professional emergency management, the mechanism by which government discharges this perceived obligation.

Emergency management rests on three pillars: a knowledge of history, an understanding of human nature expressed in the social sciences, and specialized technical expertise in response mechanisms.

History tells what happened, suggesting what events could occur again, and provides examples of how others have dealt with crisis. Social science suggests why people react to crisis in certain ways and why some methods of crisis response succeed and others fail. The technical expertise demanded of the emergency manager addresses how crisis is managed, both in the immediate response, but more importantly, in the development of strategies to reduce risk and build community resilience. This chapter considers the first leg of the tripod on which emergency management rests, that of historical perspective.

WHY STUDY HISTORY?

In a recent posting to the email list service for the International Association for Emergency Managers, a veteran emergency manager asked the question: "Given our advances in technology, what can we possibly learn from past disasters?" His query was prompted by a student's request for information on bioterrorism response and the suggestion by several members that the student examine several historical disasters, particularly the Flu Pandemic of 1918. The emergency manager asked the question as to how an early twentieth century response using limited technology could have relevance for a practitioner in the twenty-first century, where mass communication is almost instantaneous and modern methods of treatment and diagnosis are so much further advanced. It is a reasonable question.

A slightly different way of phrasing the question is, "Does the study of the history of disasters and the development of response mechanisms have any relevance for emergency managers?" The answer is a resounding, "Yes!"

It is common to look at an institution, process, or system and assume that it sprang into existence as the result of conscious decision-making. In other words, there is an assumption that the end result is what was intended all along. In reality, institutions tend to develop and evolve over time and are influenced by a variety of factors. The result is almost always an imperfect system that is a result of compromises between an ideal desired state and political and economic realities, many of which shift over time. Understanding how an institution

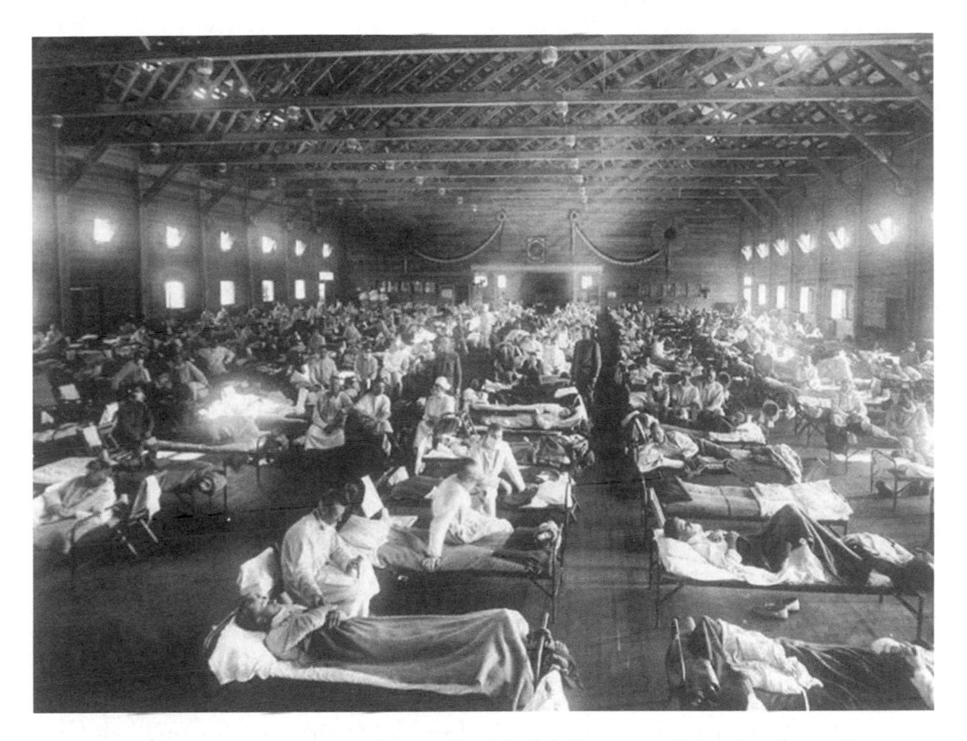

Figure 1.1 Emergency hospital during the 1918 influenza pandemic, Camp Funston, Kansas. *Courtesy of the National Museum of Health and Medicine, Armed Forces Institute of Pathology, Washington, D.C.*

developed can tell a lot about why it operates the way it does and the likelihood of changing or adapting to future conditions.

The study of historical disasters and the evolution of response systems offer a number of benefits to the emergency manager.

- Historical disasters suggest what could happen in a community. In conducting a hazard analysis, many planners are driven by the immediate past or by disasters that had a major impact on the community. For example, emergency planners in San Francisco are heavily influenced by the memory of the 1906 earthquake and fire and by the 1989 Loma Prieta earthquake. On the other hand, New Yorkers plan almost exclusively for terrorist attack, a legacy of the

September 11, 2001 tragedy. Few San Franciscans are aware that hazards such as tsunamis and tornadoes have occurred in the history of their city or that an outbreak of the plague occurred in the early 1900s. Likewise, many New Yorkers would be surprised to find that a fault line runs under Manhattan Island.

- Historical disasters identify hazards that are foreseeable. Emergency planners are expected to identify and plan for hazards that could impact the community. Events that are foreseeable are of particular concern as the question is always asked after the event "Why didn't we plan for this?" Prior to September 11, 2001, would anyone have thought that an airplane could be used to attack a building? Possibly not, but the possibility of an airplane flying into a building was foreseeable. On July 28, 1945, a military bomber whose pilot was disoriented by fog and drizzle crashed into the Empire State Building, igniting a major fire and forcing the evacuation of 1,500 people. This historical incident did in fact influence building design throughout the United States. These design parameters helped prevent the immediate collapse of the World Trade Center buildings and allowed the safe evacuation of thousands of occupants.
- Historical disasters provide long-range views of potential hazards. By nature, people tend to focus on the near term and on incidents that are part of institutional memory. However, a brief review of the historical record, particularly those events involving geological hazards, can give one pause. Prior to the 1990s, few remembered the series of earthquakes that shook New Madrid, Missouri, in 1811-1812. Yet these were among the largest earthquakes in history, changed the course of the Mississippi River, destroyed almost 150,000 acres of forest, and were felt throughout most of the United States. The overall impact was low because of the sparse population of the region at the time. The probability of a similar earthquake is estimated at 90 percent by 2040 and would result in great loss of life and billions in damages.
- Historical disasters give an indication of the social impacts of disasters. When most people think of disaster, they think in terms of immediate needs and the response to the consequences of the event. An emergency manager, however, must be concerned with

the long-range impact and the need to restore the community to some semblance of normalcy. By studying the long-range impact of historical disasters, it may be possible to make reasonable assumptions as to what might occur. While the Black Death of the 1300s is remembered for its toll on human lives, it had a profound impact on the economics and social contracts of the Middle Ages, leading ultimately to increased wages, better living conditions, and the end of the feudal system. In his recent book, *Krakatoa: The Day the World Exploded: August 27, 1883*, author Simon Winchester traces the rise of Islamic fundamentalism in the Eastern Pacific as a direct result of the eruption.

- Historical disasters provide examples of cascading events. Cascading events are disasters in which an initial incident creates a ripple effect that triggers additional disasters or expands to encompass areas that would otherwise be unaffected by the incident. A recent such event was the meltdown at the Chernobyl nuclear power plant in 1986, where the spread of fallout affected the three Soviet states of Ukraine, Belarus, and Russia and reached as far as Scandinavia and Europe. The historical record is replete with many such events. In April 1815, Mount Tambora, an island east of Java, exploded, spewing an estimated 1.7 million tons of ash and debris into the atmosphere and casting a blanket of ash over one million square miles. The lighter ash formed a barrier at an altitude of some 25 miles that reflected sunlight back into space, resulting in a cold wave in 1816 that sparked worldwide crop failure and famine. A similar record of unseasonable cold and famine found in Dark Age chronicles has also been associated with a volcanic event in the Pacific.

A BRIEF SURVEY OF CATASTROPHE

The gift shop at FEMA's Emergency Management Institute in Emmitsburg, Maryland used to offer a t-shirt with a cartoon of Noah's Ark and the caption "The Original Emergency Manager." One can certainly make a case for this: Noah conducted a hazard analysis based on a reliable weather forecast and the potential for catastrophic loss of life and developed and implemented plans to mitigate the effects of the coming calamity (selection of the animals) and to respond to it (the construction of the ark). At the onset of the event, he implemented his plan, conducted a pre-planned evacuation, and managed a sustained shelter and feeding operation for 40 days and 40 nights. At the end of the incident, he implemented a recovery plan by releasing his animals and re-establishing a community.

Whether or not Noah is the original emergency manager, there is no question that as long as there have been disasters, people have attempted to avoid them and deal with the consequences. One could make the argument that primitive attempts to touch and influence the Divine through sacrifice and ritual are in fact an early form of mitigation, an attempt to avoid disaster or reduce its impact by placating some outside force. The myths and legends that predate recorded history contain many examples of attempts to stave off or prepare for catastrophe, many of which may be inspired by actual historical incidents. Stories of a great flood are common to all cultures and variations on the story of Noah can be found in the Sumerian tales of Gilgamesh, in the Indian stories of Vishnu, and even in the *Popul Vuh* of the Mayans. Ancient texts also contain examples of more prosaic attempts to mitigate catastrophe: the *Book of Genesis* tells of Joseph's stockpiling of grain against the predicted years of famine in Egypt.

These early legends of catastrophe are more than just simple stories. In many cases, archeological evidence suggests that they might in fact be records of actual historical disasters. Geologists Walter Pitman and William Ryan in their book, *Noah's Flood: the New Scientific Discoveries about the Event That Changed History*, theorize that Noah's flood may recall an event in 5600 B.C.E. that led to the formation of the Black Sea. According to the theory, the sudden formation of the Bosporus Strait caused the Mediterranean Sea to flood into an existing freshwater lake at the rate of ten cubic miles of water per day (about two hundred times the rate of Niagara Falls), creating what is now known as the Black Sea. Plato's writings on Atlantis in 350 B.C.E., while generally regarded as myth, may in fact be related to the volcanic destruction of the island of Thera circa 1628 B.C.E. that may have led to the end of the Minoan civilization in the Aegean.

Not all ancient disasters are accessible only as myths. Historians of all ages have evinced an interest in disaster and those of the Classical

Age are no different. Many of these accounts are factual records based on first hand observation. Thucydides provides a vivid description of the plague that struck Athens 430-428 B.C.E. and presaged the end of the Golden Age of Greece. Thucydides' record is of particular interest as he contracted and survived the plague.

Perhaps the most gripping eyewitness account of catastrophe is Pliny the Elder's description of the eruption of Mount Vesuvius and the destruction of Pompeii and Herculaneum in 62 B.C.E. While a minor footnote in history now, modern scientists have estimated that the eruption of Mount Vesuvius released thermal energy equivalent to 100,000 times that of the atomic bomb dropped on Hiroshima, Japan in 1945. The consummate naturalist, Pliny chose to leave a safe vantage point in Misenum and move closer to Pompeii to make more detailed observations from the deck of a small galley. He recorded his observations moment by moment until he collapsed and died, most likely from a heart attack brought on by overexertion.

Accounts of catastrophe such as those of Thucydides and Pliny offer more than just historical interest. In many cases, they contain detailed observations of the impact of the onset and effects of the event. These accounts also chronicle the reaction of people affected by disaster and allow one to draw conclusions about human nature. The reaction of people to catastrophe in the past is not significantly different from those in the present time.

As one considers the historical record, it is also possible to identify precursors to modern emergency management concepts. Over time, one begins to note the demand by people that their governments provide them with protection against potential disasters, leading to striking parallels to current emergency management concepts and methods. This suggests that there may be a certain commonality in how humans respond to crisis. This gives rise to the speculation that there may be certain default mechanisms for crisis response that are both natural and comfortable. If so, the closer emergency plans resemble this mechanism, the more effective they are likely to be.

Elements of emergency management make their appearance early in recorded history. In response to public demand that he do something about the fires that frequently plagued Rome, the Emperor Augustus established a brigade of vigiles in 6 C.E. that served as fire fighters and night watchmen. The brigade consisted of seven cohorts of 560

Figure 1.2 Mount Vesuvius from space.
Courtesy of the National Aeronautics and Space Administration.

men each, with each cohort responsible for two of Rome's 14 districts. A special detail of 320 vigiles was drawn from the Roman cohorts and rotated to the ports of Ostia and Portus every four months. The vigiles were stationed in seven major barracks and 14 watchtowers and performed nightly patrols to ensure safe use of fire and lamps. In addition to the expected hooks, axes, ladders, rope, and buckets, equipment included high pressure water pumps capable of reaching to a height of

Figure 1.3 Thirteenth Century illustration showing people covered in the buboes, characteristic of the Black Death. *Toggenburg Bible 1411.*

60 to 100 feet. Each cohort also had four physicians and the Roman equivalent of a chaplain (victimarius) assigned to it. The parallels to modern fire fighting organizations are striking.

In the plague that struck Europe in 1348 one can see reflections of modern fears of bioterrorism or pandemics, such as SARS or Asian bird flu, and of modern emergency management practices. The arrival of the disease was stunningly swift: within three years, anywhere from 25 to 50 percent of Europe's population had been killed. Major cities such as Florence, Venice, Hamburg, and Bremen lost at least 60 percent of their populations. Paris lost 50,000 people, half of its population. Many of the problems and solutions found in the plague record are reminiscent of those faced by twenty-first century responders. To deal with the vast amount of sick people, makeshift hospitals, usually run by the clergy or volunteers, were established. These makeshift hospitals would have looked remarkably similar to the temporary hospitals of 1918 shown in Figure 1.1. The large number of dead almost immediately overwhelmed the churches that normally disposed of remains through burial in consecrated ground, forcing the use of mass graves and cremation, techniques that are still used in disasters despite the fact that such practices are unnecessary and create severe problems for the victims' families. Cremation and mass burial continue to be used for the same reasons they were used in the Middle Ages: fear of contagion and a misunderstanding of the health hazards posed by mass fatalities.

The Black Death forced a reexamination of traditional medical practices. Up to this point, many treatments were based on biblical cures and traditional remedies. The almost universal failure of these remedies resulted in their replacement by techniques based on observed results. In some cases, observation supported traditional methods. For example, the frequent misdiagnosis of the plague as leprosy led to the imposition of a biblically mandated 40-day quarantine. Despite this misdiagnosis, physicians noted that quarantining those ill with plague helped protect others. The success of this practice led in turn to harsh quarantine restrictions in several areas. The city of Pistoia ceased all imports of wool and linen and barred citizens who had visited plague-ridden towns from reentering the city. In other locations, such as Milan, houses where the plague appeared were sealed up, with both the dead and living inside. Other jurisdictions used fire to burn out the contagion. These measures did result in slowing the spread of the disease.

The Severe Acute Respiratory Syndrome (SARS) in 2003 forced a reexamination of the use of quarantine as a method for preventing the spread of contagion. The epidemic began in November 2002 and came to public attention in February 2003. In April, the World Health Organization (WHO)issued an advisory recommending only essential travel to Toronto, Canada based on the number of cases there and concerns over the potential spread of the disease to other countries. However, only one confirmed case of SARS has been "exported" from Canada and new cases were all from among medical personnel who had treated SARS patients and who were being subjected to severe quarantine restrictions. Nevertheless, the WHO persisted in prompting similar advisories from other countries that produced a severe blow to tourism in Canada. In the reaction of the WHO, one hears echoes from Pistoia.

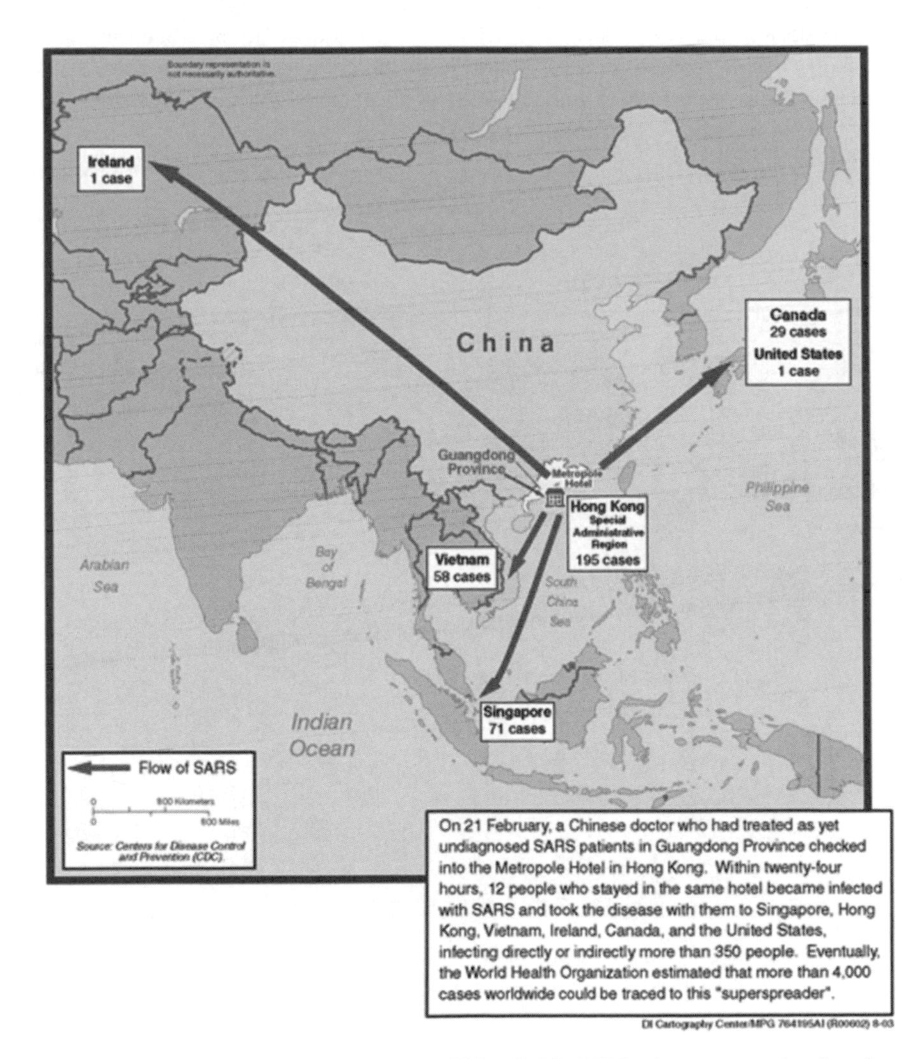

Figure 1.4 Spread of SARS virus as of March 28, 2003, about a month after the initial outbreak in February. *Courtesy of the Central Intelligence Agency.*

CASE STUDY: THE LISBON EARTHQUAKE OF 1755—COORDINATED DISASTER RESPONSE

The first "modern" disaster is considered to be the Lisbon earthquake of 1755. At 9:20 on the morning of November 1, Lisbon, Portugal was struck by an earthquake estimated to be close to 9.0 on the Richter scale that lasted anywhere from three to six minutes. The trembler spawned a tsunami that struck the city shortly after in three waves. Fires started in the areas not inundated by the tsunami and burned for five days. Eighty five percent of Lisbon's buildings were destroyed and as many as 90,000 citizens are thought to have been killed, roughly one third of the city's population. In all, the earthquake and tsunamis claimed between 60,000 to 100,000 lives worldwide and the shockwaves were felt as far away as Finland and North Africa.

(continued on next page)

Case Figure 1.1 Lisbon, Portugal was virtually destroyed on November 1, 1755 by an earthquake, tsunami and fire. Copper engraving dated 1755.

(continued from previous page)

The response to the Lisbon earthquake is the first example of a coordinated government relief effort. The Prime Minister, Sebastião de Melo, Marquis of Pombal, took charge and directed the chief justice to appoint 12 district leaders with emergency powers. He ordered the Portuguese army to surround the city to prevent workers from leaving and to provide transportation for the delivery of food from outside the city. Food prices were controlled to prevent gouging. Fishing was encouraged and taxes were suspended for fish sold in the impacted area. Workers were then pressed into service in debris clearance and body recovery. Contrary to the customs of the time, Pombal directed the disposal of remains by mass burial at sea. Gallows were erected to deter looters and at least 34 such looters were executed.

As the situation came under control, Pombal turned to reconstruction. Almost immediately he ordered military engineers to draw up new plans for a redesigned city. He ordered a survey of property rights and claims and passed legislation to prevent evictions and to control rents. Debris was sorted to salvage building materials. Most importantly, Pombal studied the effects of earthquakes on building design and incorporated the results into the reconstruction, creating the first seismically protected structures in the world and giving birth to the science of seismology.

Pombal's response to the Lisbon earthquake is significant because he used an approach that not only addressed immediate issues but also looked toward the future. It was not enough, for example, to feed the hungry—by using economic incentives in suspending the tax on fish he encouraged the development of a new industry and potential tax base. Pombal had a clear vision—prior to the earthquake he had been dedicated to modernizing Portugal and increasing its influence throughout the world. He viewed the earthquake as a mere delay in achieving that goal and focused his efforts on restoring the city of Lisbon as quickly as possible. He used the opportunity presented by the earthquake to build a better, safer city.

The relationship between disaster and attempts to prevent its reoccurrence or reduce its impact is one that repeats itself throughout history. The Great Fire of London in 1666 offers another example of this relationship. Started by a careless baker who had failed to douse his oven fire at night, the fire burned for five days and destroyed some 436 acres, most within the City of London. Eighty-seven churches and

13,200 houses were destroyed, about 90 percent of the city's houses, and between 100,000 to 200,000 people were left homeless. Oddly, only six fatalities are recorded, although there is anecdotal evidence that the initial toll was much higher and increased over time as survivors succumbed to disease and starvation. (Inaccurate casualty figures are common in historical accounts: until 2005, the official death toll for the 1906 San Francisco earthquake and fire was only 478. Research by historian Gladys Hansen convinced the Board of Supervisors to amend the official count to over 3,400. The original number was deliberately distorted for political reasons.)

Following the fire, houses were rebuilt from brick and stone instead of wood and thatch to reduce the risk of fire. Insurance companies were chartered to provide both insurance indemnification and, in some cases, fire fighting companies that would attempt to rescue insured properties. A water main system of hollowed tree trunks was laid under city streets

Figure 1.5 London, as it appeared from Bankside, Southwark, during the Great Fire—from a print of the period by Nicholas Visscher. Robert Chambers, *Book of Days*, 1864.

to provide water to fire fighters, who would dig down to the main and drill a hole to access the water. (The wooden plug used to seal the hole is the origin of the term "fireplug" still in use today.)

These examples of historical disasters demonstrate a number of key points. First, they emphasize that while catastrophic events often seem sudden and unexpected; in many cases they are at least foreseeable. This unwillingness to see the possibility of catastrophe and to instead treat foreseeable disasters as "Acts of God" is a common theme throughout history. The refusal of the Lord Mayor to deal with London's fire risk in 1666 is echoed again in the failure of San Francisco's Mayor Schmitz to act on the warnings of his fire chief prior to the earthquake and fire in 1906. The potential for failure of the New Orleans levee system was well known prior to Hurricane Katrina, but little was done to prevent it.

A study of historical disaster shows another disturbing trend: the practice of preparing to deal with the last catastrophe instead of considering all possible hazards. The Great Plague of London had begun in 1664 and reached its peak in 1665, killing 68,596 (this is the official figure but it is highly likely the total exceeded 100,000). The primary concern of the citizens of London, despite warnings and predictions of destruction by fire that included a letter of concern from King Charles II to the Lord Mayor, was plague, not fire. One sees the same pattern in the focus on nuclear war planning in the Civil Defense Era that resulted in inadequate response during Hurricane Hugo in 1989 and Hurricane Andrew in 1992. The parallels to the overwhelming emphasis on preventing terrorism since September 11 at the expense of all-hazards planning and the resulting debacle during Hurricane Katrina are readily apparent.

A subtler lesson, however, is how quickly good emergency management practices can deteriorate. London had a fire-fighting capability similar to that established by Augustus in Rome in the first century. However, when the legions withdrew in C.E. 451, the system was allowed to fall into disuse. William the Conqueror established a law in the eleventh century requiring that all fires be put out at night to reduce the danger from fire (called couvre-feu or "cover fire" from which we derive the modern term "curfew") but this was not well enforced in 1666. Building codes that dated from the twelfth century requiring party walls to be built of stone were likewise unheeded.

There are, again, similar parallels to San Francisco in 1906. The city had suffered from several major fires throughout its brief history and had established a series of underground cisterns to provide fire fighters with emergency water sources, the first being built in 1852. By 1866, a system of some 50 cisterns was in place across the city. The last cistern was built in 1872, at which time the city had 64 cisterns with a capacity of three million gallons. The advent of volunteer fire departments and the development of steam pumps made the system obsolete in the public mind and cisterns were routinely used to hold rubbish or were destroyed during construction. By 1906, only 23 cisterns were still listed as active, the rest being considered "lost." Pleas by fire chiefs for funding for the system, including Dennis Sullivan in 1905, went unheeded, with catastrophic results.

History also holds the lessons of those who got it right. The first "modern" disaster is considered to be the Lisbon earthquake of 1755. The earthquake and the tsunami that followed were the eighteenth century equivalent of the Indian Ocean earthquake and tsunami on December 26, 2004. The response by the Marquis of Pombal was effective because it was based on a long-range strategy that incorporated reconstruction, economic incentives for growth, and mitigation against future disasters.

The Lisbon earthquake is also significant because it marks the beginning of a change of attitude toward the nature of disaster itself. For much of recorded history, disasters have been viewed as "Acts of God." Viewed this way, disasters are beyond control and nothing can be done to prevent them from happening. However, the Lisbon earthquake occurred in the middle of the Age of Enlightenment, at a time when reason was valued over traditional beliefs. In this context, there was an attempt to define disasters in scientific terms and to view them as "Acts of Nature," implying that disasters could be both explained and affected by human action.

It is also interesting to note that one of the reasons the Lisbon recovery was so effective was that it fit in with the long-term vision of the community. The Marquis of Pombal had been working for years to make Portugal a significant world power. The earthquake afforded an opportunity for advancing his vision through a program of modernization during reconstruction. This integration of community vision into

CASE STUDY: THE TEXAS CITY DISASTER—GOVERNMENT NEGLIGENCE

Shortly before 8:00 a.m. on April 16, 1947, a fire was discovered in the hold of the Grandcamp, a former mothballed Liberty ship that had been provided by the United States government to assist in Europe's reconstruction. The vessel was loaded with some 7,700 tons of ammonium nitrate, along with some small arms ammunition and other goods, and was loading in Texas City, Texas, because the Port of Houston did not allow the loading of such dangerous cargo.

Efforts to extinguish the fire were ineffective, in part because the French captain did not want to ruin his cargo by dousing it with water. Instead, he ordered the hatches closed and covered with tarpaulin, turned off the ventilators, and turned on the steam system. It was only when the heat forced the crew off the ship that the fire department was finally notified.

At approximately 9:12 a.m., the Grandcamp exploded, hurling huge metal fragments, some weighing several tons, which sparked numerous fires as they ripped through buildings and oil storage tanks in nearby refineries. The blast was said to have knocked people off their feet in Galveston, 10 miles away, and shattered windows in Houston, 40 miles away. The shockwave was felt in Louisiana, some 250 miles away.

Moored at the adjoining slip to the Grandcamp was another Liberty Ship, the Grandflyer, loaded with 900 tons of ammonium nitrate and 1,800 tons of sulfur. The explosion on the Grandcamp set this ship ablaze, forcing the crew to evacuate after an hour of firefighting. The Grandflyer exploded at about midnight, adding to the carnage.

The Texas City explosion is considered the worst industrial disaster in United States history. The official death toll was 581, with an additional 113 missing, and over 5,000 injured. Over 500 homes were destroyed, displacing some 2,000 people. Damage was estimated at over $600 million (in 1947 dollars).

The explosion spawned hundreds of lawsuits against the federal government. Many of these were consolidated in a class action suit, *Elizabeth Dalehite, et al. v. United States.* On April 13, 1950, the district court found for the plaintiffs, citing numerous acts of omission and commission on the part of 168 government agencies that resulted in the explosions. The Fifth U.S. Circuit Court reversed the decision on the basis that the district court had no jurisdiction to find the government as a whole liable for the negligence of subordinate agencies. The Federal Tort Claims Act, the legislation under which the suit was brought, exempts "failure to exercise or perform a discretionary function or duty" and the court found that all the acts of commission or omission were discretionary in nature. The Supreme Court, in a four to three opinion, affirmed the ruling in 1953.

(continued on next page)

(continued from previous page)

The decision in *Elizabeth Dalehite, et al. v. United States* is significant because it affirmed the protections of government agencies against personal responsibility. There was no question that the government was negligent. Justice Robert Jackson noted this in his minority opinion, "...the disaster was caused by forces set in motion by the government, completely controlled or controllable by it. Its causative factors were far beyond the knowledge or control of the victims." However, the fear was that a victory for the plaintiffs would paralyze government and open the door for many more lawsuits of a similar nature. The need to protect the government's ability to act, to exercise "discretion," was paramount. Justice Stanley Reed alluded to this in his majority opinion, "Congress exercised care to protect the government from claims, however negligently caused, that affected governmental functions."

reconstruction is a precursor of what in now known as holistic disaster recovery.

The study of historical disasters is also a study of the record of progress in public protection. Tragically, it sometimes takes a catastrophe for people and government to make the social changes necessary for effective mitigation and response. As was noted, the Great Fire of London spurred the development of improved fire prevention methods and led to the growth of fire insurance. San Francisco today has a system of cisterns that are regularly maintained. Smaller events likewise played a role: the Triangle Shirtwaist Factory fire in New York on March 25, 1911 led to the development of new safety laws for factories in the State of New York, with some 36 new laws being passed between 1911 and 1914.

Not all disasters lead to progress. Following the Texas City disaster on April 16, 1947, the survivors sued the United States government for negligence in preventing the explosion of two ships loaded with ammonium nitrate bound for farmers in Europe. The case was the first class action lawsuit against the United States government and went all the way to the Supreme Court. Although a lower court had found for the plaintiffs, the Fifth Circuit Court of Appeals and the Supreme Court determined that the government was not liable for negligent planning decisions on the part of subordinate agencies, thus establishing a presumption of limited governmental liability that continues to this day.

MODERN EMERGENCY MANAGEMENT EVOLVES

As one studies the history of disaster, one soon notices a trend over time for governments to become increasingly involved in emergency management. Hence one sees rulers such as Augustus and William I establishing rudimentary fire codes and committing public resources to enforce them. Following the Black Death, towns in Italy in the 1350s started to implement initiatives aimed at controlling public sanitation. In London, following the Great Fire, the government issued charters to insurance companies to support fire suppression efforts. However, these measures were almost exclusively based on a disaster that had occurred or specific to a regularly occurring hazard.

This tendency of government to be reactive to disaster crossed the Atlantic to the New World. In December 1802, the town of Portsmouth, New Hampshire, suffered a catastrophic fire that burned 132 buildings and caused the fledgling U.S. Congress to pass an act in 1803 to make federal resources available for assistance, the first national disaster legislation in the United States. (Portsmouth would suffer several more "Great Fires," including one in 1813 that destroyed the entire central business district—244 buildings.) Congress would continue this pattern of providing assistance through special acts, passing some 128 such pieces of legislation between 1803 and 1950.

The executive branch of government also began to be involved in the provision of relief. In 1900, President William McKinley issued a charter to the American Red Cross (which is still in effect today) that established the Red Cross' role in disaster relief. This charter would have its first significant impact just six years later when President Theodore Roosevelt directed the Red Cross to coordinate relief efforts following the 1906 San Francisco earthquake and fire. President Franklin D. Roosevelt enhanced this more active role for the federal government as part of the New Deal package of legislation, providing loans for the repair of public structures damaged by disasters and grants for repairing federal roads and bridges. Among this legislation was the Flood Control Act of 1936 that was aimed at reducing vulnerability to flood across the country.

The early twentieth century also saw the government taking a more proactive role in involving the public in national security issues. The advent of war in 1916 led to the United States Army Appropriation Act of 1917 that established an embryonic civil defense structure consisting of the Council of National Defense (CND) and a network of state and local councils. These councils would languish both during and after World War I, but as hostilities grew in the late 1930s, the councils were resurrected and the CND was linked to a new Office of Emergency Management created in 1939 by President Franklin D. Roosevelt. In 1941, President Roosevelt abolished the CND and created an Office of Civil Defense (OCD) within the Office of Emergency Planning. President Harry Truman, in turn, abolished the OCD following the cessation of hostilities in 1945.

Following the dissolution of the OCD in 1945, the government commissioned a number of studies on civil defense that urged evacuation planning and shelter construction to protect the civilian population from the nuclear threat of the Cold War. In 1948, President Truman created an Office of Civil Defense Planning and in 1949 temporarily assigned civil defense responsibility to the National Security Resources Board before finally creating the Federal Civil Defense Administration in December 1949.

This piecemeal approach to dealing with disasters was to change significantly in 1950 when two important pieces of legislation laid the foundation for modern emergency management in the United States. The Federal Disaster Act of 1950 established the federal government's role in domestic disaster relief while the Civil Defense Act of 1950 established a federal, state, and local framework for preparedness. It is important to note that, at this point, the two functions of relief and preparedness were not viewed as components of a system but rather as two separate functions. For almost 40 years, the two functions would develop along separate lines.

The Federal Disaster Act of 1950 had its genesis in the need for an ongoing authority to respond to disaster. As has been noted, between

1803 and 1950, Congress had enacted 128 separate acts to provide relief after local disasters and larger catastrophic events. The process was cumbersome and meant that most local disasters did not receive federal aid. The Federal Disaster Act provided for a continuing authority for the federal government to provide assistance without returning to congress for separate legislation for each disaster.

Unfortunately, Congress continued its reactive approach to disasters. Programs were created or expanded by legislation passed in reaction to significant events, creating a confusing fragmentation in relief programs. Many of the programs and structures now considered an integral part of federal relief, such as the Federal Coordinating Officer and the Public and Individual Assistance Programs, were created in this manner. An attempt was made to reduce fragmentation with the Disaster Relief Act of 1970 and again in 1974 with the passage of the Robert T. Stafford Disaster Relief and Assistance Act (the Disaster Assistance Act of 1974). The Stafford Act also included, for the first time, the provision for federal assistance for disaster preparedness and warning programs.

The Civil Defense Act of 1950 defined the role of local government in disaster preparedness. The act had its genesis in the growing fear of nuclear war and the beginning of the Cold War. The Soviets had detonated a nuclear bomb in 1949 and in 1950 North Korea invaded South Korea. Consequently, the primary focus of the act was on national security. The Civil Defense Act of 1950 stated that the principal responsibility for civil defense rested with state and local government and gave authority to the FCDA to provide guidance and resources to assist in this process. The act was amended in 1958, in response to state and local government complaints, to provide for joint responsibility for civil defense among all levels of government.

Almost immediately, the federal government's emphasis on national security met with resistance. Local governments were more concerned with natural disasters that occurred with some regularity than with the potential for nuclear war. This tension led eventually to an insistence on the use of resources provided under the Civil Defense Act for planning for all hazards, not just nuclear war. The act was amended in 1976 to allow the use of civil defense resources for disaster relief and recovery from natural disasters (referred to as the Dual Use Doctrine) and again in 1981 to allow dual use of government-provided resources for preparedness.

Along the way, responsibility for civil defense shifted through a confusing array of offices that migrated from the Office of the President to the Department of Defense and back to the White House. A similar process was taking place with the responsibility for disaster relief and the congressional approach of reacting to the needs of current disasters resulted in a wide range of relief programs scattered throughout various departments and agencies. In 1979, President Jimmy Carter consolidated these scattered programs and merged the functions of preparedness and response into a single agency, the Federal Emergency Management Agency (FEMA).

The result was almost immediate chaos. The new agency was composed of programs transferred from a variety of other agencies, each with a staff with its own corporate culture and with a separate funding source and congressional oversight committee. For the first year, staff continued to operate out of separate offices scattered around Washington D.C. and it would be over 10 years before the agency became truly effective. Nevertheless, for the first time, the functions of preparedness and disaster relief were the responsibility of a single agency. This consolidation of function received a boost in 1988 with a major amendment to the Stafford Act that effectively merged the Civil Defense Act and the Disaster Relief Act. For the first time, the United States placed responsibility for preparedness and relief operations in a single agency with a single legislative authority.

The formation of FEMA did more than consolidate various federal relief and preparedness programs. The Civil Defense Act had spawned a network of civil defense offices in each state and in many counties and local jurisdictions. Initially, these positions were funded at a 50 percent match and were limited to work on national security projects such as shelter surveys and crisis relocation planning. As the Dual Use Doctrine became more acceptable to the federal government, many of these planners began to focus on planning for and responding to natural disasters. FEMA became a source of policy guidance and a perceived repository of expertise. Through its grant requirements and the establishment of a national training center, FEMA also had the potential to standardize emergency management training and doctrine across the United States.

Nowhere was this leadership role more apparent than in the tenure of James Lee Witt as Director from 1993 to 2001. Up until Director Witt's nomination by President Bill Clinton, FEMA had been led by a succession of political appointees with limited experience in emergency management. Witt was the first director to have served as head of a state emergency management agency. Shortly after his arrival, Witt issued a set of agency goals and a strategic plan, the first time this had ever been done in FEMA. His principal goal was not to increase response capacity but instead "to make mitigation the cornerstone of emergency management." Witt practiced what he preached, lobbying for increased post-disaster mitigation funds under the Stafford Act and ultimately for pre-disaster funding of mitigation projects. This emphasis sent a clear message to the emergency management community—emergency management is not just about response.

A major impetus behind the formation of FEMA was a 1978 report from the National Governors' Association concluding a study of emergency management policies and practices in the United States. The report cited the fragmentation of operations, the lack of connection to state policy, and the lack of an integrated national policy or strategy. The report went on to espouse a new concept called comprehensive emergency management that took an all-hazards approach to dealing with risk and emphasized interagency cooperation. The report also delineated a four-phase model of emergency management, combining preparedness, response, recovery, and mitigation into an interrelated process.

Comprehensive emergency management rapidly became the basis of a national strategy for emergency management and was quickly adopted by the newly created FEMA. The model was broad enough to encompass both the national security requirements of the federal government and the natural hazard concerns of local government. This broad scope meant that emergency planners now had to consider all the potential hazards of a jurisdiction and not just specific planning requirements such as crisis relocation planning. As emergency planners became more educated across a broad spectrum of subjects, a specialized body of technical knowledge related to the practice of emergency management began to emerge.

The embryonic profession of emergency management took a major step forward when the National Coordinating Council on Emergency Management (NCCEM) (now the International Association of

Figure 1.6 Comprehensive Emergency Management Model.
Courtesy of the Federal Emergency Management Agency.

Emergency Managers—IAEM) established the Certified Emergency Manager (CEM) designation in 1990. The CEM designation established a baseline of knowledge and experience for members of the profession. The CEM program will be discussed in more detail in Chapter 3.

In 1991, NCCEM, FEMA, and the National Emergency Managers Association (NEMA) helped to establish a Disaster Management Committee to develop standards for emergency management programs. The work of the committee was released under the auspices the National Fire Protection Association (NFPA) first as a recommended practice and then, in 2000, as an American National Standard, NFPA 1600 Standard on Disaster/Emergency Management and Business Continuity Programs. NFPA 1600 initially received little attention in the emergency management community. Following the events of September 11, there was an increased demand for standards to guide emergency management programs and the 9-11 Commission Report recommended the adoption of NFPA 1600 as the National Preparedness Standard. This was done in 2004 as part of the National Intelligence Reform Act.

NFPA 1600 was used as the basis for the development of a program to assess jurisdictions against the standard. This program, the Emergency Management Accreditation Program or EMAP, was a joint project of FEMA, IAEM, and NEMA to adapt the general NFPA 1600 standard specifically to state and local government. The candidate jurisdiction performs a self-assessment against the EMAP Standard and the results are peer reviewed by an assessment team of emergency managers. The assessment team provides input to an accreditation committee that makes the final determination on accreditation.

While NFPA 1600 and the EMAP Standard are voluntary, they are making progress toward being generally accepted in the emergency management community. This was greatly assisted by FEMA's contracting with the EMAP Commission to develop a baseline assessment of state programs in 2005 under the National Emergency Management Baseline Capability Assessment Program (NEMB-CAP) and the recommendation in the Homeland Security Grant Guidance that states moving toward adoption of the EMAP Standard. The federal fiscal year 2006 guidance for the Emergency Management Performance Grant requires that states use the EMAP Standard and the results of NEMB-CAP as a basis for developing work plans and performance evaluations.

THE IMPACT OF HOMELAND SECURITY

For many, the attacks of September 11, 2001, are considered the genesis of the concept of Homeland Security. The U.S. concern over terrorism began that day and required new approaches and new mechanisms of national security. Almost immediately, over 50 years of evolution in emergency management were casually dismissed with the stock phrase "the world has changed."

To emergency managers, however, the truth is somewhat different. Few outside the emergency management community noticed that in February 2001 the Stafford Act was amended to include terrorism and other catastrophic events in the definition of major disaster. This is significant in that it allows the provisions of the Stafford Act to be used in response to such events. (The Stafford Act is very specific and invoking it sometimes requires feats of bureaucratic legerdemain—in the aftermath of the riot following the acquittal of the officers involved in the Rodney King beating in Los Angeles in 1992, FEMA technically responded to the destruction caused by the fires rather than to the riot itself. The Stafford Act at the time covered fires but excluded civil disturbances.) Even fewer members of the public noticed the introduction in March 2001 of HR1158 that sought to establish a National Homeland Security Agency or HR1292 that required the development of a Homeland Security strategy.

Prior to September 2001, national security had always been of concern to emergency managers, both because of their roots in the Civil Defense program and because national security issues are considered in the all-hazards strategy of comprehensive emergency management. Emergency planners had worked on nuclear war planning and had been involved in terrorism planning, most notably in the development of the Metropolitan Medical Response System fostered by the Nunn-Lugar-Domenici legislation of 1996. Emergency planners had also helped develop a field operating structure, the Incident Command System, that could be used to coordinate the response activities of multiple agencies and had helped formulate the Federal Response Plan, the plan that coordinated overall federal response to disasters.

Yet, suddenly, everything was "new" and the emphasis on terrorism, coupled with a large influx of federal funds, brought new players to the game, many of whom had little to no knowledge of comprehensive emergency management. Terrorism response was viewed as primarily a law enforcement function and much of the initial funding was devoted to building a response capacity for first responders. In many jurisdictions, emergency managers were sidelined by new Homeland Security offices and there was growing concern that emergency management grants, on which many emergency management offices depended, would be reduced or eliminated.

Part of this marginalization stemmed from a decision by James Lee Witt in 1997 to distance FEMA from terrorism planning in favor of mitigation by turning down the opportunity to start what would become the Office of Domestic Preparedness (ODP). Witt's rational was that the proposed program was narrow in scope (i.e. aimed primarily at the law enforcement community) and would draw agency resources from FEMA's priority of creating community resilience through mitigation. ODP was placed in the Department of Justice and, over time, built a strong constituency among law enforcement officials. In the bureaucratic infighting during the formation of the Department of Homeland Security (DHS), ODP acquired FEMA's grant programs, stripping staff and funding from the ailing agency.

FEMA even began to lose credibility among emergency planners on the federal level. Stripped of its cabinet rank, the agency was not even tasked to revise the Federal Response Plan that had been one of its major accomplishments. Instead of tweaking the proven plan, Secretary Tom Ridge sought a completely new approach and tasked the Transportation Security Administration to develop the new National Response Plan that was used in Hurricane Katrina. The result of this marginalization was that many experienced FEMA staff left the agency to take jobs in state or local government or in the private sector. The combination of this flight of expertise and an unproven national plan was a recipe for disaster in Hurricane Katrina.

The adverse impact of Homeland Security has not been felt solely at the national level, but has caused considerable disruption with local emergency management offices. In viewing the lessons of September 11, federal officials have focused on initial response rather than on the full range of social, political, and economic impacts of the event. This in turn leads to emphasis that is effectively operational in nature rather than strategic.

Consider, for example, the issue of interoperability. Homeland Security planners have responded to the failure of communications at Ground Zero by looking at technological solutions. Get everyone on the same radio system and the problem will be solved. However, even a cursory reading of after action reports from the incident reveals that the problem was not technological in nature—New York fire and police personnel traditionally do communicate with each other because of a mutual animosity that has nothing to do with technology. Further, New York did not use an incident management system and had no centralized capability to manage the incident. This is in stark contrast to the scene at the Pentagon where responders from Arlington

County immediately established a unified command, implemented a mutual aid communications plan and coordinated the response of multiple local, state and federal agencies. Interoperability, in the tactical sense, equals radio systems, but strategic interoperability means the ability of disparate organizations to operate together under a common plan.

This same confusion over goals was evident in the initial Homeland Security grants. Funding was provided primarily for the purchase of protective equipment for local responders. However, no funding was available for training personnel in how to use the equipment. Worse, no funds were provided for the planning that would have determined needs and leveraged existing capabilities. Providing individual protective equipment is operational in nature. A strategic approach would have taken into account the poor state of existing capability and looked for ways to increase overall capacity. The Council on Foreign Relations 2003 report on the use of Homeland Security grants, *Emergency Responders: Drastically Underfunded, Dangerously Unprepared*, noted that basic capabilities to respond to structural collapse, detect hazardous materials, or identify biological agents were at extremely low levels and had not been significantly enhanced by the grant programs.

The Council report also noted that it was difficult to determine critical needs because there were no standards that represented a minimum acceptable level of capacity. Barred from using grant funds for enhancing current capability, jurisdictions had no way to measure the difference between existing and desired capacity and so funds were often used to purchase equipment that would, in the long run, prove to be unnecessary.

Homeland Security's focus on the operational has resulted in a commitment to command and control methodologies reflected in the National Incident Management System (NIMS). While NIMS is based on the highly regarded Incident Command System, ICS is primarily a field operating system that is useful for hierarchical paramilitary organizations. In other words, it works for fire, police, and emergency medical services. NIMS fails to take into account, however, the qualitative differences that emerge as one approaches crises of increasing complexity that must be managed by non-hierarchical organizations.

The methodology used for coordinating response as opposed to commanding response is the Multi-Agency Coordination System (MACS). The 152-page NIMS guidance document devotes less than three pages to MACS.

The "down in the weeds" mindset of DHS is further amplified in a complex guidance document, the Target Capabilities List (TLC). The TCL "defines the capabilities, outcomes, measures, and risk-based target levels of capability for the nation to achieve" the National Preparedness Goal. In essence, the TCL is a series of checklists of critical tasks that must be performed to achieve a desired outcome. Here again, however, the emphasis is on the tactical—the TCL covers chemical, biological, radiological, nuclear, and explosive attacks in detail but devotes less than 20 of its 509 pages to the critical tasks that are the primary concerns of local government: damage assessment, restoration of lifelines, and community and economic recovery.

This blind spot in Homeland Security is unfortunate. NIMS represents a major step forward for emergency management. A common operating system enhances interoperability and makes the provision of mutual aid considerably simpler. The ICS principles on which NIMS is based are proven in both the emergency management and business communities. The TCL is a step toward providing the standards and metrics called for in the Council on Foreign Relations study. There is also no question that at least some of the Homeland Security grant funds have been put to good use to enhance local capacity to respond to multiple events. However, the single-minded focus on terrorism response, as reflected in the guidance documents and grant requirements issued by DHS, represents a significant step backward from the concept of comprehensive emergency management.

THE IMPACT OF HURRICANE KATRINA

There is no doubt that Hurricane Katrina in August 2005 represents a major turning point for emergency management in the United States at least on a par with September 11. Katrina demonstrated with uncompromising harshness how far the level of preparedness in the United States had been allowed to decay since September 11, despite the billions in funding provided under Homeland Security programs under the theory that "if you're prepared for terrorism, you're prepared for anything." The fact is that first responders are not equipped to cope with large-scale disasters and catastrophe. For crises of these proportions, there must be a community-level response supported by state and federal resources. Coordination and not command becomes paramount and it is the non-paramilitary agencies such as public health and public works and voluntary agencies like the Red Cross and Salvation Army who must work together to restore the community.

However, it is too easy to blame the failures in Hurricane Katrina solely on the failure of government. The dynamics of a catastrophe are too complex. Instead, it is useful to ask, "By what standards have we decided that government failed?" Catastrophes and disasters by nature stretch response capabilities. They are overwhelming. They are always confusing and frequently mishandled. Mistakes are made, some of which turn out to be major. New research into the 1906 earthquake and fire in San Francisco, for example, suggests that much of the fire damage could have been prevented if the decision to use explosives to create firebreaks had not been made. The use of black powder as a blasting agent by untrained responders actually spread the fire.

There were certainly some successes in Katrina. The 80 percent evacuation rate of the New Orleans area is a higher percentage than usually expected in evacuations. The National Guard had a significant presence in the area within three days instead of the five required in Hurricane Andrew. By September 1, 2005, 20,000 Guardsmen were deployed in Louisiana, Alabama, and Florida. By the same date, FEMA had deployed 1,800 specialized personnel such as National Disaster Medical Assistance Teams and urban search and rescue teams. The Coast Guard had deployed 4,000 Coast Guardsmen and rescued close to 2,000 people. The U.S. Department of Agriculture delivered 80,000 pounds of food to support shelter and feeding operations. This was supplemented by 13.4 million liters of water shipped by the Department of Transportation. DOT had also shipped 10,000 tarps, 3.4 million pounds of ice, and 144 generators to the affected area. The Red Cross was caring for nearly 46,000 evacuees in over 230 shelters. Leaving aside the question of adequacy, the response to Hurricane Katrina represents the largest and most extensive relief operation in U.S. history.

At the sites of the operation's biggest failures, the New Orleans Superdome and the Ernest N. Morial Convention Center, conditions were, by all accounts, horrific. But the violence and anti-social behavior described in early reports could not be substantiated. Ten bodies were recovered from the Superdome and four from the convention center, only one of whom was considered to have died under suspicious circumstances. While numerous people claimed knowledge of rapes and murders, officials could find no one who had actually witnessed the crime and no victims came forward. Further, if one considers that Hurricane Katrina struck New Orleans on Monday, August 29, and the Superdome and convention center were evacuated by Friday, September 1, one could argue that the response was close to the 72 hours that disaster officials maintain is the minimum time one can expect to be without services following a disaster event.

This is not to suggest that the response to Katrina should be considered adequate or that the deplorable conditions at the Superdome and convention center could, under any circumstances, be thought justified or acceptable. Rather, it points out how public expectations can color the perceived success or failure of response operations.

In some sense, emergency managers at all levels are victims of their own success. The professionalism of the first responder community, coupled with the overwhelming response brought to bear by FEMA in smaller disasters, has created a public perception that the government has inexhaustible resources that can be brought to bear immediately in any crisis. This has been a natural outgrowth of the historical trends this chapter has traced.

Prior to Hurricanes Hugo and Andrew, FEMA was viewed as a third tier responder. The agency generally arrived on scene a week or two after a disaster and handed out checks to assist with reconstruction. The public perception of failure in Hugo and Andrew significantly changed FEMA's role to that of more immediate response. This, in turn, forced changes to the Stafford Act authorities and agency procedures and led ultimately to the development of the Federal Response Plan. FEMA suddenly found itself with coordination responsibility for direct federal response.

Hurricane Katrina will force similar changes on DHS. There are already calls to restore FEMA to an independent agency status with cabinet rank or to abolish it in favor of a new agency, the National Preparedness and Response Authority. The National Response Plan is under revision. Local and state plans are under review, with emphasis on evacuation planning, special needs populations, and logistics systems.

However, the inexperience of the DHS planners continues to show. Rumors abound that the next revision of the National Response Plan will include another layer of management similar to the discredited Principal Federal Official position held by Michael Brown. The failure of this position in Katrina, and the similar failure of a "direct representative of the president" with no authority in the 1992 Los Angeles riots, have not yet sunk in. Further, the DHS push for evacuation planning uses a hurricane model not suitable for most jurisdictions where the need for evacuation will be more immediate than in a hurricane. In this sense, DHS seems doomed to repeat the mistake of the Crisis Relocation Planning program of the Cold War era: failing to understand the unique needs of local governments.

CONCLUSION

The confused response of DHS to the failure of Hurricane Katrina relief operations highlights the need for an understanding of history. The increasing role of government in disaster response has been directly attributable to the demand for protection from the public. The success of government in meeting this need has led to increasingly higher public expectations. In failing to manage those expectations by admitting that there are things that government cannot or should not do, political leaders are betting that they will not be caught out by a major disaster that will expose the gap between public expectation and reality.

It is not difficult to draw historical comparisons between Homeland Security and the early days of the Cold War. There is the same vague, but potentially devastating, threat from an external agent and the same uncertainty creating an overwhelming perception of risk in the general public that is out of proportion to the realities of individual risk. By separating emergency planning into "terrorism and everything else," the United States has in essence stepped back over 50 years to a bifurcated strategy that is driven by perceived risk rather than a dispassionate risk analysis and capacity building.

When one examines this from a historical perspective, what has been taking place in the United States since 2001 is really nothing new. Periodically through history, there have been national security concerns that have engaged public interest with resulting congressional attention. In many cases, these have resulted in increased mandates for local government, usually without funding and without any real commitment of new resources to all-hazards planning. This is particularly true in the emergency management community where many offices are scrambling to deal with Homeland Security without any additional staff or funding. However, this historical perspective suggests that "this too shall pass" and that the pendulum will return to comprehensive emergency management as legislative attention is drawn to new crises. The corollary to this is that, armed with this long-range perspective, the emergency management community may be able to take advantage of the current situation to make progress toward the goal of comprehensive emergency management.

Emergency management stands at a crossroads. Over the centuries it has evolved from isolated reactive measures based on a specific disaster to an established function of government. This government function has itself evolved from a reactive approach to a full strategic concept. At the same time, practitioners have developed whose specialized expertise has led to the emergence of a new profession with a code of conduct, professional certifications and standards, and a specialized body of knowledge. The current emphasis on Homeland Security would seem to have been a major setback for comprehensive emergency management, but taken from a historical perspective, it will most likely force a further evolution in the role of the emergency management practitioner.

Chapter 2

EMERGENCY MANAGEMENT: A SOCIAL SCIENCE PERSPECTIVE

This is a disaster. This isn't something somebody can control. We ain't stuck on stupid.
—Lieutenant General Russel L. Honoré
Commander, Joint Task Force Katrina

The historical study of disasters provides a fertile source of information for emergency managers. However, this information is often anecdotal and incomplete. A great deal of study is required to identify trends and draw conclusions that are useful in emergency planning. To meet this need, emergency managers must turn to the discipline of social science.

Social science is the study of human society and of individual relationships in, and to, society. Where history tells what occurred, social scientists attempt to determine why something happened. From this, it is possible to predict what is likely to occur in the future and to base emergency planning on realistic possibilities rather than on the many myths and unchallenged assumptions that are common to emergency planning.

SOCIAL SCIENCE AS AN EMERGENCY MANAGEMENT TOOL

For too many emergency managers, social science is an overlooked tool. In fairness, there are reasons for this. Academics tend to write for other academics, sometimes making their work difficult to read for the layperson. Researchers tend to focus on empirical studies that can be very limited in scope and not directly relevant to emergency managers. Few researchers are interested in describing the broad conclusions that might be useful for emergency planners. Pure research is favored over applied science. Researchers are viewed as "ivory tower academics" with no practical experience in emergency response operations and therefore lacking in credibility. This attitude completely ignores a discipline that is constantly evolving and increasing in utility and that is building the theoretical knowledge base necessary for emergency management's evolution as a profession.

Part of this problem lies with the emergency management community itself. As was noted in Chapter 1, the discipline of emergency management evolved from the civil defense programs of the 1950s. This evolution put a premium on previous military or emergency services experience, disciplines that have little connection with academic studies related to emergency management. Many emergency managers are retired from these services and are not even aware of the vast body of social science literature available to them.

Like the discipline of emergency management itself, social science research into disasters began in the early 1950s during the Cold War. The U.S. government sponsored research with the goal of anticipating how people would react in a nuclear attack by using disasters as a type of "natural laboratory." The first large scale studies were conducted by the University of Chicago's National Opinion Research Center between 1950 and 1954. Interestingly, these studies were sparked by an actual occurrence—a hazardous materials incident involving sulfur dioxide fumes in Donora, Pennsylvania in October 1948. Military officers sent to study the incident hoped to draw conclusions about civilian behavior in a poison gas attack. Instead, they were puzzled when they observed unaffected people mimicking the symptoms of the victims (what is now known as the worried-well syndrome) and sought an explanation for the phenomena through further study. Other studies followed, most notably the work of the Disaster Research Group of the National Academy of Sciences in the mid-1950s.

While the initial intent of these studies was to study individual reaction to wartime situations, a major lesson learned was the need to study the reaction of organizations. While the studies did not produce many results that were particularly relevant to wartime needs, they did suggest that many of the common assumptions about individual and group behavior under crisis were false. These popular "disaster myths" are discussed in more detail below.

An interesting phenomenon about this initial research is the manner in which sociologists co-opted purely military projects to explore human reaction to crisis outside the scope of warfare. The work of the Disaster Research Group is a case in point. Despite being established to study disasters caused by enemy action, the day-to-day work was dominated by sociologists, resulting in a basic research orientation rather than one of applied science. In other words, the results of the studies provided valuable information for the overall understanding of disasters, but provided few practical applications that could be applied to wartime issues.

Of significant interest to the emergency management community was the establishment of the Disaster Research Center (DRC) at Ohio State University in 1963. A proposal by Ohio State's Sociology Department to study organizations under stress came at the same time that the Office of Civil Defense was both concerned over civilian behavior in war (the Cuban Missile Crisis had occurred in October 1962 and the Cold War was heating up) and had the funding to support research. The original proposal was expanded at the request of the Office of Civil Defense and the Disaster Research Center was established. Moved to the University of Delaware in 1985, the DRC conducts field and survey research into community-wide crises and serves as a repository for materials collected by other agencies and researchers. Its library of over 50,000 items and series of over 400 books, monographs and reports are available to agencies involved in emergency management, as well as to researchers.

A counterpart to the DRC is the Natural Hazards Research and Applications Information Center (NHRAIC) at the University of Colorado at Boulder, founded in 1976. While its mission is similar to that of the DRC (i.e. to serve as a clearinghouse of knowledge on the social science and policy aspects of disasters), the NHRAIC focuses primarily on natural hazards and mitigation. In addition to a library and publications series, the NHRAIC publishes a twice-monthly newsletter for developers and users of hazard information and a bimonthly periodical dealing with current disaster issues. The NHRAIC also has a quick response program funded by the National Science Foundation that funds social science research in the immediate aftermath of a disaster to capture perishable information.

The evolution of social science research into disasters has special significance for the emergency manager on several levels. Social science research offers insights into the nature of disasters, but, more importantly, addresses the issues of public expectation. As discussed in the previous chapter, public expectation is both a key driver behind the evolution of emergency management and an indicator of success. Understanding how people react to crisis and what methodologies are most likely to work is critical for an emergency manager.

On another level, emergency management is evolving as a profession. One of the hallmarks of a profession is a specialized and theoretical body of knowledge. For much of its history, the emergency management discipline has considered this specialized body of knowledge to be the skills related to emergency response, such as the ability to establish shelters or to conduct evacuations. However, this is the knowledge required of a technician, not a professional, and in many cases, these actions can be, or are, performed primarily by other disciplines. The theoretical knowledge that forms the basis of emergency management lies not in these technical skills but in social science research and a deeper understanding of the nature of disaster and the reaction of people and organizations to crisis.

EMERGENCIES, DISASTERS, AND CATASTROPHES

What, exactly, constitutes a disaster? Over the years, the word "disaster" has come to mean many things to many people. In most cases, it's used as a generic term for "something bad has happened," but surely in a book that deals with emergency management one can be more precise. The task is not as simple as one might think. The first question that arises is "Whose definition should be used?" There is the definition from *Merriam-Webster's Collegiate Dictionary, eleventh edition*, which, under the proposed 2007 changes to NFPA 1600 may be included in the National Preparedness Standard. There are the definitions provided by various agencies, such as the Federal Emergency Management Agency (FEMA), and there are definitions used under various pieces of legislation such as the Stafford Act. Each of these definitions has been formulated for a specific purpose and has a slightly different meaning. Some of these definitions will be discussed later in this chapter.

Social science researchers have argued that disasters are social constructions. That is, that they are defined by the nature of their impact on social systems. An event occurring where there is no population does not usually rise to the level of a disaster unless it produces cascading effects that have an impact on society. There is also evidence to suggest that disasters are a manifestation of vulnerabilities in modern society, an argument that certainly correlates well with the definition of risk discussed in a later chapter. One must also consider that, as society grows more complex, conflicts between human activity and natural processes may have unintended consequences.

Defining what is meant by disaster is not just an academic exercise–it is a reflection of how society deals with the consequences of crisis. As environmental historian Ted Steinberg notes in *Acts of God: The Unnatural History of Natural Disaster in America*, the assumption that disasters are random events beyond the control of society has been used as an excuse for a lack of mitigation efforts and failure to plan on the part of government and the general public. If one accepts, however, that disasters are a direct result of the vulnerability of society to specific events, most of which are foreseeable, then it follows that realistic steps can be taken to reduce vulnerability and to prepare for the consequences of disaster.

What truly complicates the issue of defining disaster, however, is that the impacts of disasters are ultimately relative. That is, they are functions of the nature of the disaster itself and the resources available to those who must deal with the consequences. A five-alarm fire in a major metropolitan area may be treated as a routine occurrence; the same fire in a rural setting may involve a multi-county mutual aid response. Disasters are also a function of public perception of risk. The threat of rotating power outages in California in 2001 had little impact on public safety but public concern elevated it to the level of a crisis.

Another definitional issue is that the term "disaster" may be too generic to distinguish between an event that causes significant damage and loss of life, such as a winter flood, and large-scale events with national impact, such as Hurricane Katrina. This suggests that there may be a need for a "hierarchy of disasters" to differentiate between the magnitude of events. This begs the question, "Is this difference merely one of scale?" That is, as the magnitude of events increases, are the differences only quantitative (e.g. a disaster can be responded to in the same way in which day-to-day emergencies are dealt with) or are there qualitative differences as well?

The work of social scientists, particularly that of Dr. E. L. Quarantelli, suggests that there are in fact both quantitative and qualitative differences between routine and disaster response. This supports the idea that there are distinct escalating levels of crisis that can be fitted into a hierarchical structure.

The first level of this hierarchy consists of those events that occur fairly frequently and can be dealt with internally by a community. Such events are generally referred to as "emergencies." FEMA defines an emergency as "a dangerous event that normally can be managed at the local level" (*IS 230 Principles of Emergency Management*). The Stafford Act defines an emergency for the purpose of federal relief as: "Any occasion or instance for which, in the determination of the president, federal assistance is needed to supplement state and local efforts and capabilities to save lives and to protect property and public health and safety, or to lessen or avert the threat of a catastrophe in any part of the United States." In actual practice, federal declarations of emergency are rare and limited to specific categories of federal assistance. Consequently, the IS 230 definition of an emergency as something managed by the community is sufficient for the first level of the crisis hierarchy.

In IS 230, FEMA defines a disaster as "A dangerous event that causes significant human and economic loss and demands a crisis response beyond the scope of local and state resources. Disasters are distinguished from emergencies by the greater level of response required." The Stafford Act is more specific and introduces a modifier in the term "major disaster."

> *"Major disaster" means any natural catastrophe (including any hurricane, tornado, storm, high water, wind driven (sic) water, tidal wave, tsunami, earthquake, volcanic eruption, landslide, mudslide, snowstorm, or drought) or, regardless of cause, any fire, flood, or explosion, in any part of the United States, which in the determination of the President, causes damage of sufficient severity and magnitude to warrant major disaster assistance under this act to supplement the efforts and available resources of state and local governments and disaster relief organizations in alleviating the damage, loss, hardship, or suffering caused thereby."*

In essence, FEMA's definition suggests that disasters are events that exceed local and state response resources while the Stafford Act defines disasters on the basis of impact, as determined by the president. These definitions seem to imply that the difference between emergencies and disasters is one of degree and not qualitative. That is, the difference between a major automobile accident (emergency) and September 11 (major disaster) is the number of casualties and responding resources. However, when one considers the complexity of response to larger events, the assumption that disasters differ from emergencies only in scale seems counter-intuitive.

One of the initial research issues considered by social scientists was this question of the difference between emergencies and disasters. Researchers, most notably Dr. E. L. Quarantelli, have identified at least five qualitative differences:

1. Converging organizations—emergencies are normally handled by community resources or, at most, local mutual aid. Disasters involve multiple organizations from the public and private sectors that the local community may not have worked with before. Consequently, organizations will need to be able to quickly form working relationships with different groups.
2. Loss of autonomy—disasters result in the formation of operating structures that may take precedence over normal ones. They also make use of emergency authorities that supersede property rights

such as commandeering private resources or demolishing damaged structures. These emergency authorities are usually not operative in day-to-day emergencies.

3. Performance standards—in disasters, protocols that would normally govern response are superseded (e.g. speed of response is altered, triage of patients is established, limited equipment is dispatched, etc.). This means that the level of service that citizens are used to receiving will be severely degraded. For example, despite the expectations raised by Hurricane Katrina, 911 dispatch centers are not considered a disaster response mechanism. While the communications component remains vital, the ability to receive calls and to dispatch services may be non-existent owing to damage to the telephone system and the full commitment of response resources to other tasks.
4. Closer public-private interface—disasters require the use of all available community resources and the lines between public and private resources may become blurred. This is particularly true in the area of logistics, where private sector companies are much better equipped to deliver and track critical resources. Further, the bulk of the critical infrastructure in the United States is owned and operated by the private sector, making the private sector an essential player in restoration of services.
5. Response to impact on the organization—in day-to-day emergencies, the normal mechanism of government is not generally heavily affected. In a major disaster, however, local government must provide response to citizens while dealing with the impact of the event on its own operations. Continuity plans become the key to restoring public services and the implementation of such plans may produce conflict with response plans if the two are not properly coordinated.

As one views these distinctions, it is obvious that local emergencies and major disasters are different both quantitatively and qualitatively. This implies that the command and control methodology, operational techniques, and tactics used in normal day-to-day operations are insufficient for dealing with major disasters and that organizations must do additional planning beyond that necessary for emergency response. It also implies that there may be a second hierarchal model at work in crisis, one that focuses on how emergencies and disasters are managed. These qualitative differences also tend to support the FEMA definition that disasters are the result of impacts to the responding organization that exceed available response resources.

While FEMA and the Stafford Act only define emergencies and disasters, it is clear that not all disasters are equal. Most tend to bc fairly localized events and relatively limited in scope while others have widespread effects. Using only the single term "disaster," how does one compare the scale of the tsunami that struck the Indian Ocean in

Figure 2.1 Damage in Anchorage, Alaska, following the earthquake on March 28, 1964. Measuring an astonishing 9.2 Richter, the earthquake killed 131, caused an estimated $300 million in damage and spawned a tsunami that killed 12 people and caused $15 million in damage to Crescent City, CA. *Courtesy of the United States Army.*

2004 with that which struck Crescent City, California in 1964? Are there no differences between the Alaskan earthquake of 1964 and the San Francisco earthquake and fire of 1906? Clearly, there is the need for a more descriptive term to distinguish these "super-disasters" from major disasters.

In the aftermath of Hurricane Katrina, the term "catastrophe' has emerged as the term to describe a major disaster with widespread consequences. This begs the question, though, "If there is a qualitative difference between emergency and disaster, are there differences between disaster and catastrophe?" This question was raised by the Government Accounting Office following Hurricane Andrew in 1993 and has been studied for some time by a small group of social scientists. Dr. E.L. Quarantelli has identified five ways in which a catastrophe is qualitatively different from a disaster:

1. Impact on built structures—catastrophes are characterized by heavy destruction to most or all of a community's housing stock, creating a crisis in sheltering resources. In contrast, most major disasters leave a large percentage of the community's housing stock available, allowing displaced persons to stay with friends, relatives, or volunteers. The extensive damage generated by a catastrophe may result in a scattering of the population, either through voluntary travel or through government relocation. A large portion of the New Orleans population was relocated to host communities following Hurricane Katrina; a similar relocation of population was seen after the San Francisco earthquake in 1906.
2. Impact on local officials—the normal assumption in disasters is that the local government is in charge and is in a position to coordinate relief efforts. In catastrophes, local government is disrupted either by virtue of officials being victims or by destruction to operating facilities and communications systems. This implies that the usual system of local government requesting resources may not operate in catastrophic events and that catastrophic planning at the state level should include mechanisms to push resources to affected jurisdictions.
3. Impact on assistance from local communities—the most immediate source of assistance in a major disaster is from other commu-

CASE STUDY: HURRICANE KATRINA—FAILURE OF EXECUTIVE COMMUNICATIONS IN NEW ORLEANS

An inviolate premise of federal disaster response is that the local government retains control of the disaster and that federal and state resources are deployed at the request of local government and used in support of local efforts. However, there is a growing recognition after Hurricane Katrina that this paradigm changes in catastrophe and that other levels of government must be proactive when local government cannot meet its responsibilities.

The emergency plan for New Orleans relied heavily on the continuation of basic telephone service with cell phone and satellite phone systems as backups. On August 27, 2005 the Chief Technology Officer decided to establish a command post at the Hyatt Hotel, located close to city hall because it was better served by power and food than the designated city emergency operations center (EOC). The mayor and several key city officials, including the chief of police, moved into the fourth floor conference rooms and prepared to conduct operations from that location.

Hurricane Katrina was one of the most powerful storms recorded in the Atlantic and struck New Orleans on August 29 as an extremely strong category 3 hurricane with sustained winds of 125 mph. The powerful winds tore off part of one side of the Hyatt and the storm surge breached the levee that protected the city from Lake Pontchartrain and the Mississippi River. Within hours, the Hyatt was surrounded by water.

At the Hyatt, phone service failed almost immediately because of widespread power outages and water damage to the hotel's telephone switch. The cell phone system also failed because of extensive damage to the cell sites. When attempts were made to use the satellite phones, the city team found that the batteries would not hold a charge, a fairly common occurrence when batteries are not maintained and replaced on a regular basis. The police radio system was functioning at limited capacity because of a generator failure at the main transmission site. The team was reduced to using messengers to gather information and transmit instructions.

On the evening of August 29, the team was able to establish communications with the outside world via a voice-over-IP connection on a single laptop. To expand the system, the CTO, the chief of police, and several aides drove to a nearby Office Depot and seized the equipment that was needed, including the store's computer server that was wrenched out of its brackets by the chief of police. By the evening of Wednesday, August 30, the system was operational and the mayor was communicating with the president.

(continued on next page)

(continued from previous page)

The problems faced by the mayor and his team were mirrored at the EOC. Located on the ninth floor of City Hall, the EOC also depended on basic telephone service and experienced similar problems with cell and satellite phones. In addition, the EOC lost power shortly after the storm because its emergency generator ran out of fuel.

The failure of local government communications in New Orleans had a direct impact on the provision of relief. The failure of local communications meant that information about the impacts of the storm and subsequent flooding was difficult to obtain. By isolating himself from the city's EOC, the mayor was in no position to give an accurate assessment of the damage or prioritize requests for resources. Absent this information, state and federal responders were forced to rely on other, less reliable sources of information.

nities that have not suffered from the event. In a catastrophe, large areas are impacted and local communities cannot assist each other. In addition, convergent resources such as federal assistance must be spread among many communities rather than focused on a single community as in most major disasters. This competition for resources was apparent in Hurricane Katrina where the bulk of the attention was given to New Orleans at the expense of other jurisdictions that may have had greater need.

4. Impact on community functions—catastrophic events interrupt not only the infrastructure of the affected community, but also social facilities such as schools and churches to a much greater degree than a major disaster, significantly delaying the return to normalcy.
5. Impact on political processes—because of the physical damage and the disruption of social structures, catastrophes bring to the forefront political issues that are generally overlooked during normal times. As an example, Hurricane Katrina raised issues of corruption in local government and racial stratification in New Orleans.

Quarantelli's work suggests that there is indeed a hierarchy of crisis that must be acknowledged in emergency planning and that there is both a quantitative and a qualitative difference between emergencies, major disasters, and catastrophes. The National Response Plan acknowledges this by defining catastrophic events as "any natural or manmade incident, including terrorism, that results in extraordinary levels of mass casualties, damage, or disruption severely affecting the population, infrastructure, environment, economy, national morale, and/or government functions." This is similar to the Stafford Act disaster in that it defines catastrophic events by the nature of the impact, essentially a quantitative rather than a qualitative measure.

However, the National Response Plan further clouds the issue by creating a new definition for an "Incident of National Significance." An Incident of National Significance is defined as "An actual or potential high-impact event that requires a coordinated and effective response by, and appropriate combination of, federal, state, local, tribal, nongovernmental, and/or private-sector entities in order to save lives and minimize damage, and provide the basis for long-term community recovery and mitigation activities."

There appears to be some bureaucratic hair-splitting going on here in an effort to give the director of DHS a declaration authority analogous to the FEMA director's Declaration of Disaster under the Stafford Act. One of the circumstances under which a declaration of an Incident of National Significance can be made is a major disaster or emergency as defined in the Stafford Act. The implication seems to be, particularly in light of the various reports on federal response to Hurricane Katrina, that a major disaster now requires both a declaration of disaster under the Stafford Act, and a declaration of an Incident of National Significance by the director of the Department of Homeland Security prior to the deployment of federal resources. This is a far cry from the original use of the Stafford Act where the president issued the declaration of disaster and appointed a Federal Coordinating Officer as his personal representative on the scene with full authority to coordinate all federal resources responding to the disaster.

Establishing a hierarchy of disaster that recognizes catastrophic events as different from disasters is not to say that emergencies or major disasters are insignificant events; victims do not care what level an event is considered by bureaucrats, only about its impact on them.

Nor is it intended to suggest that mechanisms used to prepare and respond to emergencies and major disasters do not have utility in a catastrophe. What this hierarchy suggests is that there are significant differences among these events that must be taken into consideration when planning for crisis.

This raises again the issue alluded to earlier about a hierarchy for responding to various levels of crisis. If there is a difference between levels of crisis, is there a difference in response as well? Common sense would suggest that there is and when one considers operational models from both the public and private sector, there certainly seems to be. Crisis response falls into three rough categories:

- Operational—operational or field response consists of the systems and resources needed to respond directly to the impacts of an event. These systems and resources include what are normally considered first responders (i.e. police, fire, emergency medical personnel) and manage response at the field level.
- Tactical—tactical response manages the overall response to the event by coordinating the activities of multiple responding agencies, anticipating resource needs, and coordinating public information.
- Strategic—strategic response manages the crisis by examining long-range implications of the event, determining long term goals and objectives, and establishing priorities that will guide operational response.

This particular model is based on a civilian business model used in corporate performance measurement but could just as easily be based on the military levels of war. The military model recognizes the three levels as strategic, operational, and tactical, which can create confusion. However, the point is that, in any disaster, there are three levels of activity taking place: a long range planning process that sets strategic goals, an intermediate level that translates strategic goals into the tactical deployment of resources, and a field component level that achieves results. These three levels correlate very well with the operational levels found within the National Incident Management System (NIMS) in Figure 2.2 and in the three-tiered gold, silver, and bronze incident management system used by the City of London (see Figure 2.3).

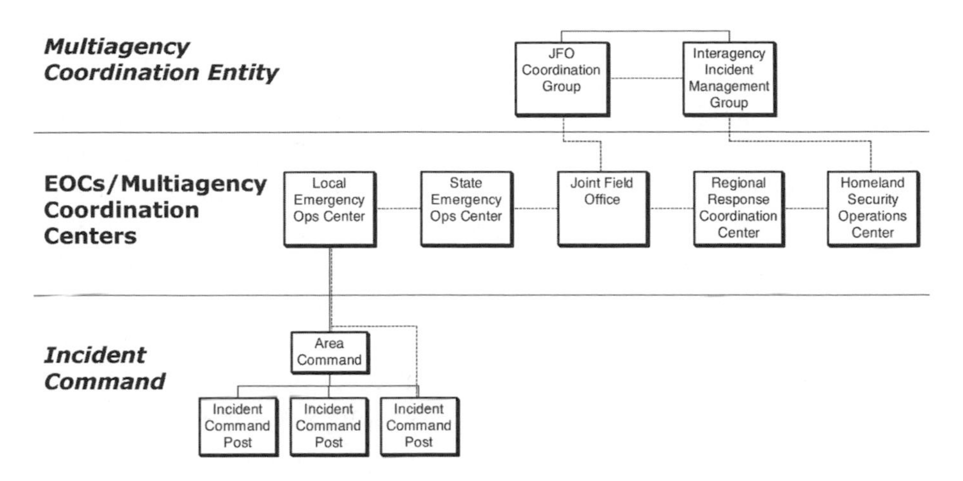

Figure 2.2 NIMS Framework National Response Plan, December 2004.

6.2 Gold, Silver and Bronze

6.2.1 'Gold', 'Silver' and 'Bronze' are titles of functions adopted by each of the emergency services and are role related not rank related. These functions are equivalent to those described as 'strategic', 'tactical' and 'operational' in other documents about emergency procedures. In summary the roles of each can be described as:

6.3 Gold (strategic)

6.3.1 Gold is the commander in overall charge of each service, responsible for formulating the strategy for the incident. Each Gold has overall command of the resources of their own organization, but delegates tactical decisions to their respective Silver(s).

6.3.2 At the outset of the incident Gold will determine the strategy and record a strategy statement. This will need to be monitored and subject to ongoing review.

6.4 Silver (tactical)

6.4.1 Silver will attend the scene, take charge and be responsible for formulating the tactics to be adopted by their service to achieve the strategy set by Gold. Silver should not become personally involved with activities close to the incident, but remain detached.

6.5 Bronze (operational)

6.5.1 Bronze will control and deploy the resources of their respective service within a geographical sector or specific role and implement the tactics defined by Silver.

6.5.2 It should be understood that the titles do not convey seniority of service or rank, but depict the function carried out by that particular person. From the outset it is important that the senior officers of each service at the scene liaise with each other. This will be the foundation upon which all later meetings will be based.

6.5.3 As the incident progresses and more resources attend the RVP, the level of supervision will increase in proportion. As senior managers arrive they will be assigned functions within the Gold, Silver and Bronze structure.

6.5.4 Senior officers arriving at their respective command/control vehicles are to establish contact with their incident commanders and should also make contact with the Police Silver in order to notify any transfer of command.

Figure 2.3 London Incident Management System. Major Incident Procedure Manual, London Emergency Liaison Panel, 2004.

This response hierarchy also correlates very well with the crisis hierarchy discussed previously. Emergencies are generally dealt with at the operational level. In the response of a community to a major disaster, the focus is primarily on tactical response. Response to catastrophes is managed at the strategic level because of the scale of the event and the long-term impact to the nation. Corresponding to each level of response is an increasing need for flexibility and creativity in the response. It is fairly easy to identify needs and resources at the tactical level. Indeed, this is done every day across the country. As one considers tactical and strategic planning, the variables begin to increase and not all situations can be foreseen, necessitating a system that is flexible enough to create response mechanisms on the fly and able to use resources creatively.

One of the results of Hurricane Katrina has been an increased level of awareness of the potential for catastrophic events. The Department of Homeland Security has launched a major assessment of local and state plans and will most likely require significant planning activities to be performed by local governments. However, at this writing, these plans show every sign of being an overreaction without any thought to the crisis response hierarchy.

To require "catastrophic planning" at the local government level shows a lack of a holistic view of emergency management. By definition, local government is fully involved at the level of disaster—there are no additional resources available that have not already been committed. If one considers Quarantelli's qualitative differences between disaster and catastrophe, local government may have even less capability in catastrophic events than exist in disasters. This means that, almost immediately, much of the responsibility for tactical response may shift to the state. This suggests that the principal planning agent for catastrophe should be the state and that local planning should be focused on building the capacity to survive by enhancing continuity planning. Figure 2.4 is an attempt to combine the levels of crisis and response and the incident management methodology of the National Incident Management System into a single model.

The response hierarchy is also operative within each level of crisis as well. For example, at the local government level, operational response is used to manage emergencies. During a major disaster, this

Level of Crisis	Definition	Principal Level of Response	Response Methodology	Principal Agent
Emergency	A dangerous event that normally can be managed at the local level	Operational	Incident Command System	Local
Major Disaster	...causes damage of sufficient severity and magnitude to warrant major disaster assistance under (the Stafford Act) ...	Tactical	Multiagency Coordination System	Local/State
Catastrophic Event	...results in extraordinary levels of mass casualties, damage, or disruption severely affecting the population, infrastructure, environment, economy, national morale, and/or government functions	Strategic	Crisis Management	State/Federal

Figure 2.4 Crisis hierarchy model.

operational response is still occurring, although, as noted, there will be changes to autonomy and performance standards. At the community level, a local emergency operations center is activated to manage tactical response, and a crisis management team of senior officials is formed to consider long-range strategic issues (see Figure 2.5).

The defining of the various levels of crisis is not just an academic exercise, it is an essential first step in understanding the nature of disaster. This understanding is necessary to ensure that plans are effective and appropriate to the needs of the community. Among other things, it determines the level at which a plan is written. This level determines planning assumptions, level of detail, and resources required. More importantly, an understanding of the dynamics of a disaster leads one inevitably to the conclusion that, as disasters become larger, the need

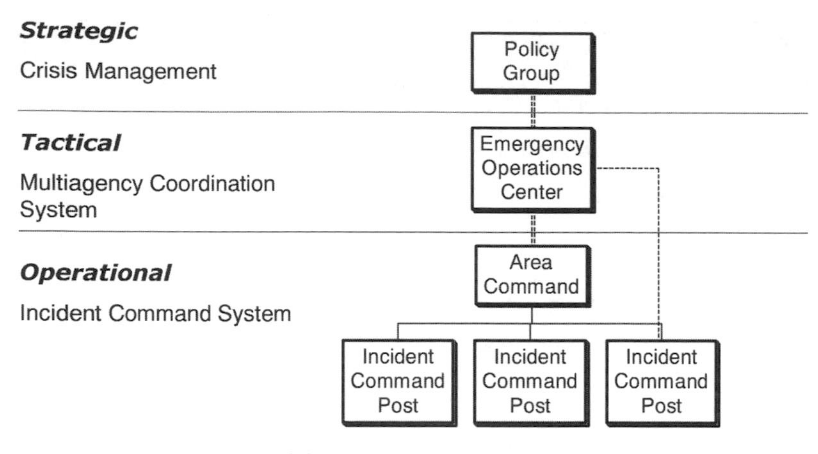

Figure 2.5 Crisis response levels in local response.

for innovation increases and plans become starting points for response, not ends in themselves. At the operational level, standard operating procedures and protocols can be used to manage the crisis. As one approaches the level of a catastrophe, plans may need to be modified or discarded and new operational structures and procedures may evolve. Pre-event planning, however, can help stimulate this process and provide a base on which to operate.

DISASTER MYTHOLOGY

Understanding the qualitative difference between emergencies and disasters and between disasters and catastrophes is a major first step in the development of realistic emergency planning. Another critical step is to understand what actually happens in disasters as opposed to what people believe will happen.

Popular "disaster" movies, unsubstantiated media stories, and "conventional wisdom" have given rise to a number of myths about how people and organizations react to crisis. Chief among these has been an assumption of antisocial behavior following the impact of an event.

People are generally considered prone to panic and to reverting to a more savage, self-centered nature, leading to a breakdown of social order and criminal activity, particularly looting. Research shows, however, that people almost consistently act just the opposite. People affected by a crisis generally become focused on loved ones and neighbors and become extremely creative in dealing with the problems generated by a disaster. Almost 90 percent of disaster victims are rescued by private individuals, not by public agencies. Panic is usually limited to situations where there is an immediate, overwhelming threat to life in a confined area and is usually of short duration.

This myth of antisocial behavior was demonstrated during Hurricane Katrina. Despite extensive media reports of looting, an investigation by sociologists from the University of Delaware showed only 237 actual arrests for looting, which, when compared to the normal crime rate, actually represented a decrease in crime from pre-disaster conditions, a situation consistent with previous research. Interviewees stated that pro-social behavior (e.g. sharing of supplies, helping neighbors) was more prevalent than antisocial behavior and suggested that many involved in what looting did occur were those likely to commit crimes pre-disaster. There was also some suggestion that some looting may have been the product of social inequality resulting in behavior normally seen in civil disturbances. While comparison with the crime rate does present problems owing to the difficulty of keeping records in a disaster and the discretion given police in determining looting versus appropriation, nevertheless it does seem as if media reports of looting were exaggerated.

Another significant myth is the impact of disasters on mental health. It is commonly believed that people are shocked into passivity, severely traumatized, and unable to care for themselves. Like the myth of antisocial behavior, there are numerous popular press accounts of people walking around like zombies following a disaster. This has led to a concern over the mental health of disaster victims and a significant commitment of resources toward mental health counseling. However, while disasters certainly produce short-term psychological reactions such as sleeplessness, loss of appetite, irritability, and anxiety, researchers have found that disasters rarely produce any new psychoses or mental illness. These psychological reactions tend to be short-lived and rarely result in behavioral dysfunction. Among a small

PROCLAMATION
BY THE MAYOR

The Federal Troops, the members of the Regular Police Force and all Special Police Officers have been authorized by me to KILL any and all persons found engaged in Looting or in the Commission of Any Other Crime.

I have directed all the Gas and Electric Lighting Co.'s not to turn on Gas or Electricity until I order them to do so. You may therefore expect the city to remain in darkness for an indefinite time.

I request all citizens to remain at home from darkness until daylight every night until order is restored.

I WARN all Citizens of the danger of fire from Damaged or Destroyed Chimneys, Broken or Leaking Gas Pipes or Fixtures, or any like cause.

E. E. SCHMITZ, Mayor

Dated, April 18, 1906.

ALTVATER PRINT, MISSION AND 22D STS.

Figure 2.6 The official proclamation issued by Mayor Eugene Schmitz following the earthquake in San Francisco in 1906. Fear of widespread looting caused him to issue the controversial order and to divert police and military resources from relief efforts. *Courtesy of the Virtual Museum of the City of San Francisco.*

number of victims, the disaster experience even seems to create a more favorable self-conception and strengthens social ties. In short, studies have shown no significant difference between pre- and post-disaster frequencies of mental health and psychological problems.

Numerous myths abound about organizational behavior as well, many of them extensions of the ones related to individual behavior. It is assumed that since workers are traumatized, organizations will be unable to function, leading to an inability to make decisions. The result of this paralysis is a deterioration of authority and social chaos. Again, research demonstrates that people and organizations begin to focus almost immediately on dealing with the consequences of the disaster and moving toward normalcy. On the morning of the San Francisco earthquake in 1906, A.P. Giannini rescued the assets of his small Bank of Italy (later the Bank of America) from the fire and within two days had established a temporary office in a private home and was giving out loans. Similarly, the day following the dropping of the atomic bomb on Hiroshima, survivors from 12 banks met together and resumed services.

The danger of believing in these myths is that they can influence the development of unrealistic plans and lead to bad decisions during a time of crisis. It is not unusual to find, after a disaster, that authorities had prior knowledge of the onset of the event but withheld warnings to "avoid a public panic." These myths can also force the deployment of resources for the wrong purpose. During the San Francisco earthquake and fire, a large number of military troops were deployed, not to perform search and rescue or assist in evacuation or fire fighting, but to prevent looting and other lawlessness, threats that never materialized. A similar situation occurred in Hurricane Katrina where troops were deployed not to deliver relief supplies but to "reestablish law and order." The real issue at the bottom of most of these myths is the assumption that people are not to be trusted in disaster and must be dealt with firmly.

This is not to say that antisocial behavior does not occur during disasters. There are certainly those within any society that will take advantage of a disruption of social order. For example, there are reports that ATMs were broken into during the confusion on September 11 and that some minor incidents of looting occurred in

San Francisco during the 1989 Loma Prieta earthquake. The issue is whether such behavior is widespread or isolated to a segment of the population that was pre-disposed to antisocial behavior prior to the disaster.

A further modifier is the nature of the event itself. Where the crisis is itself a breakdown of social order, such as in the Rodney King riots in Los Angeles in 1991, one can expect to see wide scale looting and destruction of property. The looting that occurred in the aftermath of Hurricane Hugo in St. Croix in 1989 was most likely the product of social inequality rather than the hurricane. In such events, people are heavily motivated by rage, opportunity, and peer pressure, much more so than by personal gain.

A frequent myth that comes up in dealing with response is the issue of communications failure. The inability of responding organizations to communicate with each other is frequently cited in after action reports. This failure is usually attributed to the use of multiple communications technologies that prevent interoperability. However, blaming technology for communications failure frequently ignores the fact that the organizations themselves were not communicating. Most failures that emerge from disasters are not technological in nature but rather reflect the inability of organizations to expand their internal communications structure to reflect the needs of disaster response. This failure is manifested in an inability to rapidly collect and collate information about the disaster and to communicate it with those involved in the response and the public. As crisis communications expert Art Botterell states in his Second Law of Emergency Management, "The problem is at the input," not in the technology.

Another common assumption by emergency planners is that disasters are random and affect all citizens similarly. In essence, "one size fits all." As was noted earlier, though, disasters are a product of vulnerability and factors such as socioeconomic class may determine how segments of a community are affected by a disaster. When sociological factors are brought into play, one can begin to understand adverse reactions to seemingly reasonable government requests. When dealing with evacuation, for example, research has demonstrated that people fail to evacuate not by choice but because personal circumstances such as concern over lost wages, lack of resources, or illness. A study by the University of New Orleans in 2004, a year before Hurricane Katrina, estimated that at least 100,000 residents of New Orleans had no means to evacuate during a hurricane because of a lack of economic resources.

Not all disaster myths are a result of a misunderstanding of human behavior in disaster. Many are based on information that is inaccurate but so ingrained in popular belief that it is difficult to counter. An example of this is the persistent belief that large numbers of cadavers create health hazards following a disaster. The result has been a push to quickly dispose of the dead through mass burial or cremation. Little or no consideration is given to the identification of the remains or to the cultural or religious beliefs of family members. As with most myths, this has led to a misapplication of resources that could have been used for other purposes and has increased the suffering of survivors. In the aftermath of Hurricane Mitch in Central America in 1989, concerns over contagion led to the use of scarce fuel resources for mass cremation. Following the Indian Ocean tsunami in 2004, over a thousand bodies were bulldozed into a common grave in the town of Banda Aceh, Indonesia before they could be identified.

In actuality, human remains do not create a significant health problem after disasters except under very specific circumstances (e.g. potential contamination of water sources). As body temperature falls, bacteria and viruses die quickly, preventing their transmission through vermin. Since most disaster victims are killed by trauma, there is no risk of "spontaneous" outbreaks of infectious diseases. The same is true of animal corpses as well. What this suggests is that the recovery and disposal of remains is not a first priority for emergency responders. Further, there is time to collect and identify remains and provide for disposal consistent with community values.

ORGANIZATIONAL RESPONSE

Social scientists have also assessed the response of organizations in disasters, with very interesting results. What researchers have found is that the primary source of problems in the response phase of a disaster is not the victims but the organizations attempting to help them. It is the inability of responding organizations to realize the qualitative differences between emergencies and disasters, particularly in the area of inter-organizational coordination, which leads to problems. A misunderstanding of what actually can be expected to occur in disasters leads to an overwhelming response from outside the affected area that outstrips the ability of local organizations to control it. It also leads to a reliance on insular decision making by professionals rather than the shared problem solving that proves most effective in crisis.

Part of the reason for this is what Dr. Thomas Drabek calls the "paradox of disaster." The paradox is that those most likely to respond to the initial onset of a disaster are those with the least experience in disaster response. A local jurisdiction may be affected by a single major disaster within a generation, meaning that there are few within the community with actual operational experience. State governments respond to multiple jurisdictions and so have a larger pool of experienced staff. Federal agencies such as FEMA have considerably more experience in dealing with disasters based on the frequency of response, although, as Hurricane Katrina has demonstrated, such experience is perishable and the lessons learned from previous disasters are not always the right ones. Nevertheless, the simple fact is that most jurisdictions have limited experience in managing major disasters and are not prepared to expand normal operations to encompass the qualitative changes needed to do so.

Compounding this problem is the use of a military-style model of response based on many of the myths related to disasters. The early civil defense programs were very much oriented toward nuclear war. Consequently, many of the planners hired to meet the requirements of the program were retired military personnel. Not surprisingly, these planners used the military planning techniques and assumptions with which they were familiar. The military model thus became imbedded in emergency planning and its influence continues to this day.

This is not to say that this planning model has been ineffective or wrong. It is particularly effective at the field level. (Despite vehement denials by its developers, the Incident Command System is very much based on the military staff system.) However, the system falls short of the needs of local government because it assumes that chaos will result from a disaster and that only a strong central authority (a command system) can guarantee success. Such models fail to take into consideration the fact that, unlike the military, local response is neither monolithic nor hierarchical. That is, where the military is characterized by a distinct chain of command and established control mechanisms, local response is comprised of a combination of agencies, many with their own authorities and jurisdictional concerns. Where the military can effectively employ a command and control mechanism, local government response is characterized by the need for cooperation and coordination.

As Dr. Russell Dynes points out in his landmark paper "Community Emergency Planning: False Assumptions and Inappropriate Analogies," the military model assumes that pre-emergency social organizations will collapse and that individuals will be incapable of effective personal action. Consequently, all decision making must be centralized and a new social organization must be created to compensate for the panic and breakdown of social order that follows a disaster. Further, the military model assumes that disasters are easily identified, allowing activation of the new organization to deal with the crisis. In actual practice, many disasters develop over time and may be difficult to recognize in their early stages. Dr. Dynes notes that there is a tendency to normalize events, that is, to see them as routine, and to only recognize the clues and precursors to an event in hindsight.

Crisis communications consultant Art Botterell's First Law of Emergency Management is, "Stress makes you stupid." This implies that, at times of crisis, individuals tend to operate at a level that is not conducive to complex operations. It is human nature to fall back on behavior that has proven effective in the past and represents a certain level of comfort. One can extend this to a desire to operate under the leaders and social organizations with which they are familiar. The problem with artificial constructs like NIMS and ICS is that they are attempts to create and impose a new organizational structure at the very time that individuals are seeking simplicity. If training is inadequate, as it always is, the resulting chaos can significantly hamper response. Botterell's Third Law is "No matter who you train, someone else will show up."

When one examines current emergency plans, the embedded assumptions of the military model become apparent. For example, there is a reluctance to trust conventional means of communication with the public and to control information through official releases. There is a distrust of independent actions by volunteers or emergent groups, resulting in these groups having little involvement in decision making during the crisis. There is at least a nominal expectation that personnel and organizations responding to the event will recognize and submit to a central authority. In essence, the military system is a closed system that has as its base assumptions on expectation of chaos during disasters and a distrust of individuals to make intelligent decisions about their own welfare.

As one begins to identify the underlying assumptions of the military model, it becomes apparent that they are based on disaster mythology rather than realistic assumptions about behavior in disasters. Planning based on disaster research assumes the need for coordination, not command, and for decision making that is decentralized and inclusive. Disasters create confusion and disorganization but not social chaos. People seek continuity in disasters. That is, they look to pre-disaster social structures and seek to expand or extend these structures rather than create new ones. In simpler terms, people do what they are used to doing. Consequently, effective emergency plans must integrate these existing social structures. Dynes suggests that emergency planning be couched in terms of mutual problem solving rather than as an attempt to hold social units together through the imposition of authority. In other words, replace command and control with coordination and cooperation.

A final concept that seems obvious but is often overlooked is the difference between planning for a disaster and managing a disaster. Planning puts in place the resources needed for response. It focuses on the principles that will guide response and potential tactics that can be used. In essence, planning is the development of a strategy for how the community will deal with crisis. Disaster management is the tactical and operational implementation of that strategy at the time of the crisis. It is impossible to predict how that implementation will be accomplished—the decisions that will be made, the resources that will be used, etc.—as managing involves selecting the best approach based on the nature of the crisis. How well the community responds to crisis will depend on how well the planning was accomplished and the range of options available to disaster managers.

CONCLUSION

The implications of social science research are that, for many communities, emergency plans are based on invalid assumptions about the nature of disaster. The basis for most plans is that disaster victims will need to be handled firmly and that the community can manage disasters by using a hierarchical organization and response mechanisms similar to the ones that exist in day-to-day operations. The result is an inability to cope with the changes in operating capabilities demanded by disaster and a failure to have in place an organization with the flexibility to react to the changing demands of operational response. Disasters, more than anything else, require creativity and improvisation. Paramount is the ability to rapidly assess the situation and formulate a plan of action that uses available resources in the most effective manner. The goal of the emergency management program must be to lay the groundwork necessary to ensure that the community can respond creatively and appropriately to disaster.

Chapter 3

THE EMERGENCY MANAGER: EVOLVING ROLES AND SHIFTING PARADIGMS

The single most significant societal change that has most altered community preparedness has been the increased professionalization of local emergency managers.
—Thomas E. Drabek, *Major Themes in Disaster Preparedness and Response Research*

The third leg of the emergency management tripod is technical expertise. This expertise encompasses the specialized knowledge and skills used by the emergency manager to translate the lessons of history and social science research into effective programs and plans. This chapter will consider the evolving role of the emergency manager and attempt to identify the skills and specialized technical knowledge needed in this newly-emerging profession.

CONFLICTING ROLES

Chapter 1 traced the evolution of emergency management programs from the early days of civil defense to the present emphasis on Homeland Security. In this historical survey, one gets an indication of the evolving role of the emergency manager from a focus on nuclear war planning to the full spectrum of responsibilities inherent in comprehensive emergency management. However, while one can trace the maturation of emergency management programs, defining the role of the emergency manager is more problematic.

It is astonishing to realize that there is neither a single accepted definition for emergency management nor an accepted job description for the emergency manager. Many organizations and social scientists have attempted to define this role but the simple fact is that there are significant differences across the United States in the duties and responsibilities of emergency managers.

One must turn once again to history to understand why this is so. In the civil defense era, the emphasis was on national security and emergency planners were expected to focus on issues such as crisis relocation planning and fallout shelter capacity development. Since many of these positions were part time or offered low pay, they tended to attract a large number of military retirees. As time went on and the dual-use doctrine allowed the use of civil defense personnel for planning for natural hazards, emergency management emerged as a "second career" for retirees from the military and public emergency services, primarily the fire discipline. These retirees tended to be very tactically oriented and focused on the issues inherent in planning for an effective response.

A result of this focus on tactics was the emergence of the military model of planning discussed in Chapter 2. Response planning tended to be very plan-centric and the development of multiple contingency plans or a single plan with annexes, appendices, tabs, etc. became ends in themselves. These plans were targeted to tactical issues: establishing shelters, identifying evacuation routes, providing for mass feeding, etc. In short, local disaster plans looked very similar to military plans and, since most were written in a vacuum (most military plans are written by planners, not by a group of stakeholders), they tended to be largely ignored by the community at large.

The government helped to foster this attitude by emphasizing product development as a measure of program success. Emergency management grant programs required the completion and submission of specific plans that were assessed against a template. For example, FEMA's CPG 1-8 Guide for the Development of State and Local Emergency Operations plans provided guidance for the development of emergency plans required under federal grant programs. It had a companion document, CPG 1-8A Guide for the Review of State and Local Emergency Operations, which contained a crosswalk used to determine if the jurisdiction's plan included all the elements required by CPG 1-8. This crosswalk was required to be submitted with the plan to assist in evaluating its acceptability.

The introduction of the comprehensive emergency management concept in 1978 created an immediate conflict with the status quo. Comprehensive emergency management is a strategic concept, not a tactical one, and assumes the existence of a jurisdiction-wide emergency management program. Emergency planners, having carved out a niche preparing for disaster response, lacked the knowledge base and, in most cases, the authority to implement such a strategic program.

What emerges from this conflict is a paradigm in which emergency managers are the "experts" in the four phases of the emergency management cycle from the strategic perspective but tend to specialize almost exclusively in preparedness planning. The reason for this is twofold. Preparedness planning is complex and requires a continuous effort. It can easily take up all available staff resources in an emergency management office. Secondly, it requires the detailed tactical planning that has been the hallmark of traditional emergency management. In essence, it is necessary work that coincides with a high level of comfort for emergency planners.

This emphasis on response is understandable when one considers the actual role of the emergency manager in the other three phases of the emergency management cycle. Mitigation may be the "cornerstone of emergency management" as James Lee Witt claimed, but it is very definitely strategic in nature. Effective mitigation requires full community commitment, as will be seen in a later chapter, and involves program elements such as land use planning, risk analysis, and public administration. These disciplines are specialized in their own right and well outside the experience of the typical emergency manager.

In an emergency situation, emergency managers rarely direct actual response operations. Although there are exceptions, management of a crisis usually falls to a senior elected official or to the lead emergency services agency. FEMA's Independent Study Course IS1 Emergency Manager: An Orientation to the Position likens the role of the emergency manager to that of a stage manager: the emergency manager must make sure all the resources are assembled and that the players have been rehearsed, but he or she is not the lead actor. The analogy is an appropriate one.

Like mitigation, recovery tends to be strategic, although there is a large tactical component in short term recovery (e.g. debris clearance). Not surprisingly, what recovery planning takes place is usually focused on these short-term considerations. Traditionally, emergency managers have dealt with the immediate impacts of the event and the restoration of critical services, along with initiating claims for federal reimbursement of disaster costs. Few jurisdictions have developed long-term recovery plans that consider opportunities for improving quality of life and hazard reduction.

Social science offers evidence of another drawback to expansion of the role of the emergency manager. Numerous studies have demonstrated a low interest in emergency management by many states and communities and have documented an indifference or resistance to emergency management. Public officials tend to have a limited knowledge of emergency management functions and tend to view it primarily as a response function and, consequently, the province of first responders. This attitude is evident in the location of the emergency management function in most local governments. For the most part, the emergency management function is an additional duty assigned to another agency, such as the county sheriff or the fire department. It is rare to find an emergency manager in the office of the chief elected official with the type of access one would expect for a policy advisor. This placement of the emergency management function reinforces the impression that it is a tactical rather than a strategic office.

TOWARD A DEFINITION OF EMERGENCY MANAGEMENT

Lacking a formal definition of the role of an emergency manager, one must attempt to define this role through a study of the functions of the position and the skill sets required to accomplish these functions. One can then compare the functions and skill sets of the typical emergency manager with those required by the evolving paradigm of comprehensive emergency management and the National Preparedness Standard. However, even in this relatively straightforward approach one almost immediately encounters conflicts between tactical and strategic concepts.

Figure 3.1 is taken from IS1 Emergency Manager: An Orientation to the Position, a FEMA distance-learning course intended for new emergency managers. It is a list of ten functions that FEMA defines as the major responsibilities of the position. This list has been used traditionally as a definition of the core functions of the emergency manager. To a certain extent, this is encouraged by other course materials that correctly identify mitigation as the primary responsibility of other agen-

Basic Preparedness Functions

Mobilizing emergency personnel and resources

Warning the public

Taking protective action

Caring for victims

Assessing the damage

Restoring essential public services

Informing the public

Record keeping

Planning for recovery

Coordinating emergency management activities

Figure 3.1 Basic Preparedness Functions IS1 Emergency Manager: An Orientation to the Position, FEMA.

cies and define the role of the emergency manager as "motivator, coordinator, and monitor" and "conscience of the community" but provide little additional detail. Likewise, the section on recovery is limited to short-term recovery, with only a brief mention that post-disaster mitigation is "one of your most important roles in the recovery phase."

However, if one goes back to the source material on which the list is based, *Emergency Management: Principles and Practices for Local Government*, the authors use the list to describe general functions that are needed in disaster planning in general. Their discussion focuses on the process of preparedness using a general approach, with no specific mention of the emergency manager. So while one can certainly infer that the emergency manager should have substantive knowledge of these planning functions, there is no basis for limiting the emergency manager to these functions or for making the emergency manager directly responsible for them.

Another FEMA course, IS230 Principles of Emergency Management, is not designed specifically for emergency managers but is intended to introduce the general public to an integrated emergency management system. In contrast to the list from IS1 that gives basic functions for the emergency manager, IS230 defines basic concepts for the emergency management program (see Figure 3.2). While the two lists initially appear similar, a closer reading shows an apparent contradiction. The list of basic emergency manager functions is clearly tactical—it focuses primarily on those tasks necessary for response. Even if one factors in the additional information in the course materials, the emergency manager's role appears to be primarily response oriented. The emergency management program as delineated in IS230, however, clearly addresses strategic issues in the community at large.

The list in IS230 has been generally accepted in the emergency management community as the formal definition of the functions of a comprehensive emergency management program. These same functions were used by FEMA to develop the Capability Assessment for Readiness (CAR), a precursor to EMAP, and form the basis for NFPA 1600, the National Preparedness Standard, which in turn is the basis for EMAP.

A problem with examining functional lists in the FEMA independent study courses is that it takes these lists out of context. Although the FEMA courses can be taken individually, many of them were intended

Emergency Management Program Functions

Laws and authorities

Hazard identification and risk assessment

Hazard mitigation

Resource management

Planning

Direction and control

Communication and warning

Operations and procedures

Logistics and facilities

Training

Exercises, evaluations, and corrective actions

Public Education and Information

Finance and Administration

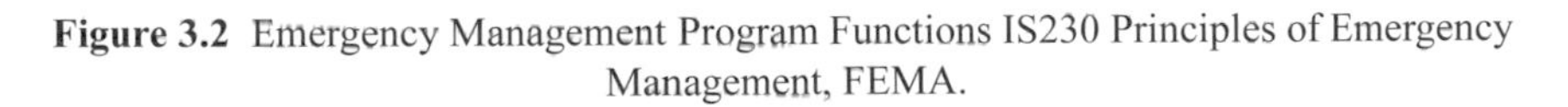

Figure 3.2 Emergency Management Program Functions IS230 Principles of Emergency Management, FEMA.

to be part of a curriculum that addressed basic skills for the emergency manager. IS 230 is a component of a Professional Development Series (PDS) of courses intended to "provide a well-rounded set of fundamentals for those in the emergency management profession." If one examines the PDS curriculum shown in Figure 3.4, the emphasis on management skills in addition to technical emergency management functions is readily apparent.

The real issue is not that there is a problem with the FEMA courses, which are excellent training tools, but rather the lack of definition of the emergency manager's role. Lacking such a definition, it is common, as demonstrated in IS1, to confuse the emergency manager with

Emergency Management Program Elements

Laws and authorities

Hazard identification and risk assessment

Hazard mitigation

Resource management

Mutual Aid

Planning

Direction, control and coordination

Communications and warning

Operations and procedures

Logistics and facilities

Training

Exercises, evaluation and corrective action

Crisis communications, public education and information

Finance and administration

Figure 3.3 Emergency management program elements, Emergency Management Accreditation Program (EMAP) Standard, Sept. 2004.

the functions of the emergency management program. The position becomes defined not by its actual requirements but by the functions of the program. It is easy to confuse the technical knowledge required of the emergency manager with the tasks that are required under the emergency management program. Yet if one examines the functions of the emergency management program closely, specialists in other disciplines perform many of these functions or have the responsibility to see that they are accomplished. In other words, while emergency managers are expected to have a basic knowledge of the functions of the emergency management program, one must, in many cases, look outside the emergency management organization to find the specialized knowledge needed to implement specific functions.

Professional Development Series

IS230 Principles of Emergency Management

IS235 Emergency Planning

IS240 Leadership and Influence

IS242 Effective Communication

IS241 Decision Making and Problem Solving

IS244 Developing and Managing Volunteers

IS139 Exercise Design

Figure 3.4 FEMA Professional Development Series curriculum.

If the emergency manager were truly defined by the functions of the emergency management program, particularly those related to response, then one would expect to see technical experts emerging as successful emergency managers. However, the research done by social scientists does not bear this out.

In his benchmark 1987 study, The Professional Emergency Manager, Dr. Thomas Drabek studied the characteristics of 12 successful emergency managers and compared them to a random sample of 50 other directors from across the country. Drabek's results were classified in three major categories:

- Professionalism—successful emergency managers had established themselves in their jurisdiction as key players through a combination of perceived expertise and contributions to overall jurisdictional goals.
- Individual Qualities—successful emergency managers shared a number of key traits such as communications skills, organizational ability, human relations skills, and control under stress. The study participants also possessed traits based on their experience that were unique to each director.

- Emergency Management Activities—successful emergency managers were committed to comprehensive emergency management, taking a broader view of their jobs than just preparedness.

The importance of Drabek's study lies in the realization that, while technical competence in the functions of the emergency management program were certainly a factor in developing professionalism, what mattered most was the perceived competence of the emergency manager as a manager and contributor to the jurisdiction he or she served. Dr. Robert Schneider in his paper, "A Strategic Overview of the "New" Emergency Management" emphasizes the importance of the emergency manager assisting in the creation of public value. In other words, to be taken seriously by elected officials, the emergency manager must be seen as contributing to larger public issues. This suggests that the role of the emergency manager is no longer that of a technocrat with highly specialized skills in emergency response but is rather that of an administrator with responsibility for overseeing the development of an enterprise-wide emergency management program.

THE EMERGENCY MANAGER AS PROGRAM MANAGER

In examining the role of the emergency manager, one is presented with a dichotomy. On the one hand there is a body of research, training materials from the premier emergency management agency, higher education curricula, and professional certifications that all point to a strategic role for the emergency manager. On the other, 50 years of actual experience and the disregard of key officials combine to limit the emergency manager's functions to the tactical arena. While it is easy to trace a historical and sociological basis for this conflict, the real problem lies in a fundamental misunderstanding of the organizational context of emergency management itself.

The single most important concept for emergency managers and the most neglected is that emergency management is a distributed process, not a discrete function. In attempting to define the emergency manager by the functions of the emergency management program, jurisdictions routinely attempt to make these functions the responsibility of a single person or office—a discrete function. No matter how well versed technically, no single person or small group can implement all the functions of an emergency management program. Instead, these functions must be distributed throughout the organization to those entities with the capability and the responsibility to perform them.

Once the concept of emergency management as a distributed process is understood, the rather indistinct role of the emergency manager becomes much more focused. An emergency manager is, first and foremost, a program manager. He or she has the responsibility for developing a strategy to guide the emergency management program and for providing oversight to ensure that the goals and objectives of that strategy are being met. This involves coordinating activities, evaluating progress, and providing technical expertise. By acknowledging this role as program manager, it is possible to now make use of the vast array of research and organizational models dealing with management.

The National Management Association recognizes four core functions of management: planning, organizing, directing, and controlling. These functions are sequential and build on each other. If the emergency manager were truly a manager, then one would expect to see these core functions contained within his or her responsibilities and this is indeed the case:

- Planning focuses on two areas: determining objectives and developing the strategies to accomplish them. This is strategic planning, not the development of emergency plans and procedures. The NFPA 1600 and the EMAP Standard require the development of a strategic plan that defines vision, mission, goals, objectives, and milestones for the emergency management program.
- Organizing defines the responsibilities and relationships of members of the organization to ensure each plays an appropriate role in achieving objectives. Emergency managers do this both at the level of the emergency operations plan that establishes functional responsibilities and operational relationships and at the level of the emergency management program when assigning responsibility for program elements within the strategic plan.
- Directing does not refer to giving orders but rather to making sure that assigned tasks are accomplished. This involves activities such as resolving conflicts, providing motivation, and evaluating performance. From the emergency management program perspec-

tive, this is accomplished through the assignment of projects to other groups in the organization and following up on project timelines.

- Controlling refers to a system of reporting, analysis, and corrective action that keeps the organizational strategy on track. This is reflected in the NFPA 1600 and EMAP requirements to assess emergency management program plans, procedures, and capabilities and to track corrective action.

This emphasis on management becomes even more evident if one looks at the emergency management program not as a discrete and separate program but as one that must ultimately add value to the organization. Traditionally, the four-phase comprehensive emergency management model of mitigation, preparedness, response, and recovery has defined emergency management with the ultimate goal being the protection of life, property, and the environment. However, one has to ask why these laudable goals and a model that has stood the test of time do not resonate with elected officials. The answer is that emergency management has always been viewed as something outside day-

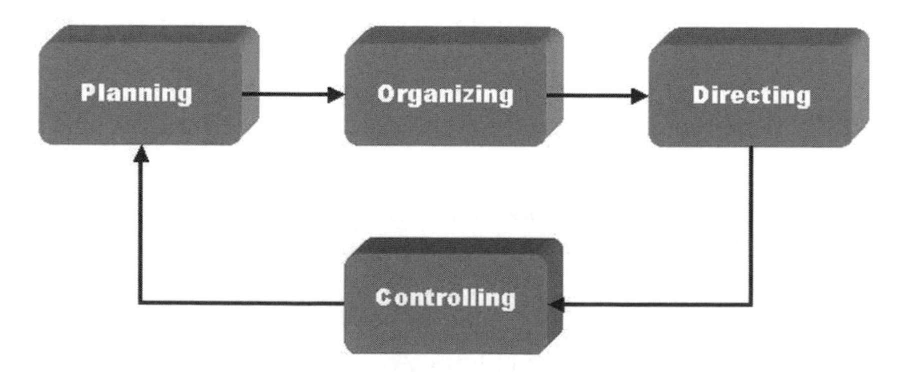

Figure 3.5 Sequential Management Functions. *The NMA Handbook for Managers*, National Management Association, 1987.

to-day government and, at best, a necessary drain on resources that could be used elsewhere.

To change perceptions, it is necessary to ask what value the emergency management program adds to a jurisdiction. If emergency management is devoted solely to dealing with emergencies and disasters, it exists only as a methodology for dealing with unexpected contingencies. In essence, it is the jurisdictional equivalent of a fire extinguisher—a necessary expense that no one expects to use. If, on the other hand, one truly accepts the concepts of comprehensive emergency management, particularly the concept of all-hazard planning, then emergency management can be defined as a mechanism by which a jurisdiction or organization manages risk.

This definition of emergency management has enormous implications. The management of risk requires a detailed hazard analysis to identify risks and the development of a strategy that determines the mix of methodologies, such as avoidance, retention, transference, and mitigation, which the organization will use to manage those risks. For those risks that are retained or cannot be fully mitigated, the organization must build a capacity to deal with the potential impact of the risk. This clearly requires a multi-disciplinary approach that must be coordinated by a generalist with a strong knowledge of emergency planning concepts. In a world of specialists, the emergency manager is the consummate generalist who must master a broad range of knowledge to facilitate the development and implementation of an organizational strategy for managing potential risks associated with identified hazards.

The acceptance of the emergency manager as a generalist does not imply that technical expertise is unnecessary. In fact, Drabek's study reinforces the need for a strong conceptual knowledge base. It merely suggests that the technical skills such as the ability to coordinate an emergency plan or conduct an exercise are core skills, not the totality of the emergency manager's art. Technical planning skills represent basic entry-level competencies that form the foundation of emergency management. However, emergency managers must move from technical specialization to strategic thinking. The entry-level coordinator knows how to develop the organization's emergency plan; the manager facilitates emergency planning for the organization. The coordinator develops an exercise; the manager oversees exercise programs.

With the role of the emergency manager defined as a program manager with responsibility for supporting the jurisdiction's management of risk, it is possible to suggest core competences for the position beyond the basic skill set expected of an entry level coordinator. Not surprisingly, these core competencies look very similar to those expected of any manager: communication (both written and oral), project management, organizational behavior, strategic planning, analytical methods, public policy, historical context, and social science research. This definition also implies that the emergency manager's technical knowledge will now emphasize strategic concepts rather than being solely task oriented.

TOWARD PROFESSIONALIZATION

The definition of the emergency manager as administrator rather than a technician seems reasonable but is the concept supportable? IS230 certainly makes the case for a programmatic approach to emergency manager. Another source supporting an enterprise-wide emergency management program is a document from the Public Entity Risk Institute (PERI), *Characteristics of Effective Emergency Management Organizational Structures*. This document does for emergency management programs what Drabek's study did for emergency managers: it identifies 20 characteristics shared by successful emergency management organizations. The document is not an actual study but is a self-assessment tool based on research conducted by the City/County Management Association in the late 1970s.

In reviewing the PERI list in Figure 3.6, one sees a correlation with the functional lists developed by FEMA, used in NFPA 1600, and the EMAP Standard. The list also contains elements of the comprehensive emergency management program: all-hazards approach, partnership, integrated emergency operations structures. All of these documents ultimately are saying the same thing: a successful emergency management program requires an enterprise-wide strategy. This emphasis on jurisdictional programs rather than discrete functions is the core of the EMAP Standard. EMAP assessors are specifically trained to look at the implementation of the emergency management program through-

Characteristics of Successful Organizations

Roles of elected officials defined

Strong and definitive lines of communications

Similar routine/disaster organizational Structures

Emergency management procedures are as close to routine operational procedures as possible

Good interpersonal relationships

Emergency management planning is an ongoing activity

All hazard approach

Disaster prevention and mitigation

Motivation provided for involvement in the emergency management program

Citizen involvement

Strong coordination among participating agencies

Public/private cooperation

Multiple use of resources

Public information function clearly defined

Ongoing monitoring for potential disasters

Internal alerting procedures

Ability to alert the public maximized

Active Intergovernmental coordination

Ability to maintain comprehensive records during a disaster

Eligibility for state and federal subsidies considered

Figure 3.6 Organizational characteristics—Characteristics of Effective Emergency Management Organizational Structures, Public Entity Risk Institute, Oct. 2001.

out the jurisdiction and to view the emergency manager as a program manager rather than the principal executive agent.

One of the primary organizations devoted to professional emergency management, the International Association of Emergency Managers (IAEM), concurs with the role of the emergency manager as program manager. IAEM administers the only recognized international certification program for emergency managers, the Certified Emergency Manager Program. The CEM certification requires that a candidate meet requirements in the areas of experience, education, and contribution to the emergency management profession. Significantly, the experience category only requires three years of actual experience while half the education requirement must be in general management subjects, supporting the idea that an emergency manager is first and foremost a manager. The candidate must also demonstrate knowledge of comprehensive emergency management through an essay and complete an examination that includes questions on general management in addition to emergency management program functions.

IAEM also offers an Associate Emergency Manager certification that requires that candidates meet the same requirements of the CEM certification with the exception of a baccalaureate degree. The implication is that the critical thinking required to obtain a baccalaureate degree is a key component of higher-level emergency management. It also suggests that emergency management as a profession may be undergoing a shift toward a two-tiered system that differentiates between those with technical expertise in core competencies (Emergency Coordinators) and those with executive skills (Emergency Managers).

Emergency management experience.* Three years by date of application. Comprehensive experience must include participation in a full-scale exercise or actual disaster.

References. Three professional references including current supervisor.

Education. Any 4-year baccalaureate degree; or additional experience may be substituted to satisfy this requirement, 2 years per 30 college credits up to the 120 credits comprising most baccalaureates. Professionals interested in recognition without the education requirement should inquire about the Associate Emergency Manager (AEM) credential.

Training.* 100 contact hours in emergency management training and 100 hours in general management training. Note: No more than 25% of hours can be in any one topic.

Contributions to the profession. Six separate contributions in areas such as professional membership, speaking, publishing articles, or serving on volunteer boards or committees and other areas beyond the scope of the emergency management job requirements.

Comprehensive emergency management essay. Real-life scenarios are provided, and response must demonstrate knowledge, skills and abilities as listed in the essay instructions.

Multiple-choice examination. Candidates sit for the 100-question exam after their initial application and the other requirements are satisfied. A pamphlet is available further describing format and sources.

* Note: A baccalaureate in emergency management reduces the experience requirement to 2 years and waives EM training if it is earned recently

Figure 3.7 Requirements for the Certified Emergency Manager® Program, International Association of Emergency Managers.

There is also some disagreement as to what the CEM certification represents. At the time of this writing, the process for obtaining certification is extremely cumbersome and the program has limited acceptance among some experienced practitioners and is not well known outside the emergency management community. Because the certification is fairly new and there are relatively few CEMs, it is difficult to make it a requirement of employment without significantly narrowing the field of potential candidates. There is also disagreement over whether the CEM certification is intended to become a minimum, entry-level certification or represents a level of expertise and experience expected of senior practitioners. Nevertheless, the CEM represents an important step for the development of emergency management as a profession and it is to be expected that the certification will evolve and adapt to the needs of the emergency management community.

The emphasis on strategic and managerial thinking is being reinforced in the training of the next generation of emergency managers. In a speech at the Emergency Management Higher Education Conference in early 2004, Dr. Wayne Blanchard, CEM, the Director of FEMA's Higher Education Project, noted that collegiate emergency management programs had grown from 95 to 113 since 2003 and 101 additional programs were under development. The Higher Education Project website listed 120 programs leading to degrees in emergency management, as of January 2006.

The Higher Education Project has developed a prototype associate degree curriculum that focuses on basic concepts and technical skills and is collaborating with academic institutions on a series of upper division courses that are clearly based on sociology and public administration. In these curricula one clearly sees the distinction between the entry-level technical expertise and managerial level strategic and conceptual thinking.

Another critical impetus toward a comprehensive program is the adoption of NFPA 1600 as the National Preparedness Standard. NFPA 1600 represents the first real attempt to establish a generally accepted standard for emergency management programs and to promote common definitions for emergency management terms. Significantly, NFPA 1600 takes a holistic view of emergency management programs and promotes the use of an enterprise-wide approach to managing risk. Intimate knowledge of the standard and an understanding of the

G230	Principles of Emergency Management
G235	Emergency Planning Course
IS-393	Introduction to Mitigation
G393	Mitigation for Emergency Managers
G275	Emergency Operations Center (EOC) Management and Operations Course
G276	Resource Management Course
G290	Basic Public Information Officers (PIO) Course
G191	Incident Command System/Emergency Operations Center (ICS/EOC) Interface
G195	Intermediate Incident Command System (ICS)
G385	Disaster Response and Recovery Operations (DRRO) Course
G202	Debris Management Course
G250.7	Workshop: Local Situation (RAPID) Assessment
G240- G242	Basic Skills in Emergency Program Management
G240	Leadership and Influence (3 days)
G241	Decision-making and Problem-Solving (1 day)
G242	Effective Communication (3 days)
G244	Developing Volunteer
G288	Donations Management Workshop
G110	Emergency Management Operations Course (EMOC) for Local Governments
G137	Exercise Program Manager/Management Course
G120	Exercise Design Course
G272	Warning Coordination
G365.3	Workshop, Partnerships for Creating and Maintaining Spotter Groups
G271	Hazardous Weather and Flood Preparedness Course
G361	Flood Fight Operations Course
G360	Hurricane Planning

Hazardous Materials Series - These courses are currently under revision and will be provided at a later date.

Figure 3.8 Prototype curriculum for associate degrees in emergency management, FEMA Higher Education Project.

requirements for compliance will ultimately form an important part of the technical expertise required of emergency managers.

A common discussion among emergency managers is whether emergency management is a profession or a technical discipline. So long as emergency managers are classed as technocrats and limited to dealing with response issues only, there is no question that emergency management will remain a discipline. However, if one accepts that the emergency manager's role is to add value to the community by coordinating a program that assists in managing community risk, then the dynamic changes.

Building Disaster Resilient Communities*
Business and Industry Crisis Management*
Coastal Hazards Management
Disaster Response Operations and Management
Earthquake Hazard Management and Operations
Emergency Management Principles and Application for Tourism, Hospitality, and Travel Management Industries*
Floodplain Hazard Management
Hazards, Disasters and U.S. Emergency Management – An Introduction
Hazards Mapping and Modeling
Hazards Risk Management
Homeland Security and Hazards
Individual and Community Disaster Education*
New Directions in Hazard Mitigation: Breaking the Disaster Cycle (Graduate)
Political and Policy Basis of Emergency Management*
Principles and Practice of Hazard Mitigation*
Public Administration and Emergency Management*
Research and Analysis Methods in Emergency Management*
Social Dimensions of Disaster*
Sociology of Disaster*
Social Vulnerability Approach to Emergency Management*
Sustainable Disaster Recovery
Technology and Emergency Management*
Terrorism and Emergency Management*

* Can be downloaded from the Higher Education web page. Other courses are under development.

Figure 3.9 Baccalaureate-level upper division emergency management courses, FEMA Higher Education Project.

While there are no standard definitions of what constitutes a profession, there are commonalities among professions. The first general requirement is for specialized knowledge. Traditionally, emergency managers have held that the list of common functions in Figure 3.1 constituted this specialized body of knowledge. However, as has been discussed, most, if not all, of these functions are actually performed by other organizations. If, however, emergency managers are something more than just technical specialists, one can make the case that emergency management theory consists of social science research, a study of historical disasters, and technical knowledge related to the standards, policies, laws, and regulations related to emergency management.

Other commonalities among professions include the existence of a professional association and a code of ethics. Emergency managers have two professional associations, the International Association of Emergency Managers (IAEM), which consists primarily of individual practitioners from the local government and private sectors, and the National Association of Emergency Managers, which is made up of state directors and high-level government officials. The two organizations work closely together and many practitioners belong to both. The International Association of Emergency Managers requires its members adhere to a code of ethics.

INTERNATIONAL ASSOCIATION OF EMERGENCY MANAGERS CODE OF ETHICS

PREAMBLE

Maintenance of public trust and confidence is central to the effectiveness of the Emergency Management Profession. The members of the International Association of Emergency Managers (IAEM) adhere to the highest standards of ethical and professional conduct. This Code of Ethics for the IAEM members and also for the Certified Emergency Managers® (whether or not they are IAEM members) reflects the spirit and proper conduct dictated by the conscience of society and commitment to the well being of all. The members of the Association conduct themselves in accordance with the basic principles of **RESPECT, COMMITMENT,** and **PROFESSIONALISM**.

ETHICS

RESPECT

Respect for supervising officials, colleagues, associates, and most importantly, for the people we serve is the standard for IAEM members. We comply with all laws and regulations applicable to our purpose and position, and responsibly and impartially apply them to all concerned. We respect fiscal resources by evaluating organizational decisions to provide the best service or product at a minimal cost without sacrificing quality.

COMMITMENT

IAEM members commit themselves to promoting decisions that engender trust and those we serve. We commit to continuous improvement by fairly administering the affairs of our positions, by fostering honest and trustworthy relationships, and by striving for impeccable accuracy and clarity in what we say or write. We commit to enhancing stewardship of resources and the caliber of service we deliver while striving to improve the quality of life in the community we serve.

PROFESSIONALISM

IAEM is an organization that actively promotes professionalism to ensure public confidence in Emergency Management. Our reputations are built on the faithful discharge of our duties. Our professionalism is founded on Education, Safety and Protection of Life and Property.

Figure 3.10 International Association of Emergency Managers' code of ethics.

A final common requirement of a profession is a process for certification or licensing. The Certified Emergency Manager (CEM) program administered by IAEM was discussed previously. The CEM designation is fairly recent and is still gaining acceptance and credibility, but taken in conjunction with other developments such as the adoption of NFPA 1600 as the National Preparedness Standard, one can certainly argue that emergency management is emerging as a profession.

CONCLUSION

This chapter has traced the dichotomy between the historical role of the emergency manager as a specialist and technician and the evolving role as administrator of the emergency management program. It has offered a definition of the emergency management program as the mechanism by which a community manages risk and suggested that such a definition offers a means of integration with community or organizational goals. In this context, the emergency manager can be defined as the administrator responsible for the strategic development and implementation of the emergency management program, a definition that implies a generalist rather than a specialist role and emphasizes managerial concepts over technical emergency management skills.

Emergency management stands at a crossroads, particularly given the advent of Homeland Security, which has created competition for the lead responsibility for coordinating disaster preparedness despite a very narrow strategic definition focused on terrorism. Social scientists agree that emergency management is evolving as a profession and a new crop of well-trained, young emergency managers are entering the profession. However, the perception of value added to an organization or community will ultimately be the deciding factor. Ultimately, the profession will need to distinguish between those who are competent in the basic technical skills needed to coordinate components of an emergency management program and those with the strategic managerial skills necessary to administer a program that supports risk management.

Chapter 4

ESTABLISHING THE EMERGENCY MANAGEMENT PROGRAM

We need emergency programs because they will demonstrate what WILL happen, with whom, when, where, and how fast. Programs will discuss relationships with other parties, and their interdependent roles. In essence, it is not about "what-if," it is about "what-is."
—Danish Ahmed, author and personal development speaker

The previous chapter made the case for the emergency manager as administrator of a program that assists the community in managing risk. This is a major change in thinking for many jurisdictions where the emergency manager has been directly responsible for tasks associated with the program. This shift of perspective also has implications for the emergency management program. It suggests that the program should no longer be viewed as the responsibility of a single individual or office but rather as an enterprise-wide program. This places responsibility for the program with the chief elected official and his or her management, with the emergency manager providing oversight to the program.

For most jurisdictions there really is no formal emergency management program. What passes for a program is usually a collection of tasks and responsibilities that have been assigned over time to the emergency manager. There has been no attempt to define expectations for the program or determine how the program can add value to the community. This chapter considers a process for establishing a formal emergency management program.

PROGRAM MANAGEMENT

As has been discussed, emergency management programs evolved over time from a narrowly defined civil defense perspective to an all-hazards approach based on the Comprehensive Emergency Management model. This growth has been through a gradual accretion of duties dictated by increased federal regulation and changing national perceptions of risk. The result has been an almost complete lack of standardization across the United States and, in most jurisdictions, programs which are little more than a conglomeration of federal grant requirements, non-funded mandates, and "other duties as assigned."

Establishing a modern emergency management program may require reassessing current activities and starting over again in the policy equivalent of zero-based budgeting. If an emergency management program is to have strategic value to the community, the process of establishing it must, in itself, be a strategic process. Formally establishing the program means to develop and document the appropriate authorities and guidance needed to administer a community-wide emergency management program. NFPA 1600 and the EMAP Standard identify four major components for establishing and managing the emergency management program:

- Program Coordinator—the community must appoint a person responsible for administering the program and keeping it current. This does not mean that the program coordinator necessarily is responsible for all program deliverables. Rather, the program coordinator helps craft strategy and provides oversight to the program.

- Advisory Committee—an emergency management program is collaborative and involves the participation of stakeholders who provide input to the program and/or assist in its coordination. This group of stakeholders provides input or assistance to the program and provides expertise, knowledge, and resources to support the program.
- Program Administration—to be effective, the emergency management program must have appropriate authority, strategy, and budget. Like any managerial program, it must identify goals and objectives and have supporting documentation.
- Program Evaluation—measures the success of the program against agreed upon performance objectives. This allows senior decision makers to determine the progress of the program and allows for identifying and fixing problem areas.

DEVELOPING A GOVERNANCE STRUCTURE

A necessary first step in establishing an emergency management program is the development of an effective organizational framework. NFPA 1600 and EMAP require appointment of a program coordinator and the establishment of an advisory committee of stakeholders that provides input or assists in coordination of the program. This advisory committee may consist of a single committee or multiple committees. However, beyond this, the standards provide no further detail.

Like everything else in emergency management, the titles, responsibilities, and authorities of the program coordinator vary from community to community. In some communities, the program coordinator reports directly to the senior executive officer, such as the mayor. In others, he or she is part of a larger entity and reports through a department head, such as the sheriff or chief administrative officer. As a general rule, the further the program coordinator is removed from the chief executive officer, the less effective the emergency management program. The program coordinator is most effective when he or she provides strategic planning and oversight and is viewed as key staff to the senior official.

The decision as to where to place responsibility for emergency management program coordination is not a simple one. In most communities, the emergency manager lacks sufficient political clout to affect positive change in their own right. Placement within the chief executive's office can offset this to some extent but may limit the resources available for program management. On the other hand, placement in an established department such as the county sheriff or an administrative department may provide access to more resources but may result in the program being viewed as that department's program with whatever political baggage goes along with that identification.

Regardless of where the emergency management coordinator is placed, the program must have an identity separate from partisan issues and be viewed as an enterprise-wide program. The program coordinator, therefore, must be viewed as a neutral party and have sufficient access to high-level management to be able to resolve conflicts within the program.

The standards require the formation of an advisory committee that provides for input from stakeholders and assists in the coordination of the emergency management program. The composition and duties of the advisory committee are at the discretion of the community. In some jurisdictions, the role of the Local Emergency Planning Committee (LEPC) required under the Superfund Amendments and Reauthorization Act (SARA) Title III has been expanded to provide oversight to the emergency management program. In others, a separate committee, such as the local disaster councils found in California, fulfills this role. The standards also allow for the use of multiple committees to oversee the program, such as a combination of a LEPC and local Citizen's Corps Council.

The composition of the advisory committee reflects the inclusiveness of the emergency management program. Where the committee consists primarily of the "Big Three" agencies (police, fire, and EMS), work products will typically reflect an emphasis on response. Where membership is broadened to include other agencies, including voluntary agencies and private sector organizations, the emphasis shifts to a more balanced implementation of the four phases of the CEM model.

The advisory committee can fulfill two very important roles in the program. First, the advisory committee provides program oversight from a broad, multi-disciplinary perspective. This helps to ensure that the emergency management program addresses the needs of the community as a whole. For example, membership on the advisory committee of advocates for special needs populations helps to ensure that these needs are at least considered in the development of a program strategy. Secondly, the advisory committee can provide resources not normally available to the emergency manager. A senior representative of a department can commit funding and personnel resources to support program strategy. Similarly, a private sector representative may be able to obtain donations of funds or resources to accomplish projects.

An advisory committee that is truly representative of the community can be very unwieldy and meeting facilitators know that the more participants in a meeting, the harder it is to make a decision. For this reason, thought should be given as to how to be as inclusive as possible in representation but at the same time avoid slowing progress. One way of approaching this problem is to distinguish between those on the committee that have actual decision making authority and those who have a stakeholder interest. Figure 4.1 shows a model that uses this approach. Program oversight and decision making is ultimately vested in a policy committee of senior department heads or elected officials. A steering committee is selected from among the stakeholders and meets fairly frequently to monitor progress and to develop the agenda for the next group meeting. The larger group would meet less frequently to review and comment on projects done by the workgroups. Completed projects are then submitted for approval to the policy committee. The advantage of this model is that it allows for broad inclusion of stakeholders in both program oversight and project implementation while retaining the ability for quick decision making.

Another useful organizational model for the emergency management program is the model used by the FIRESCOPE (Firefighting Resources of Southern California Organized for Potential Emergencies) program. The model consists of five components:

- A board of directors responsible for policy and oversight, usually consisting of agency directors;
- A coordinator who provides administrative management and is responsible to the board;
- An operations team of senior agency operations personnel that implements board decisions and makes recommendations to the board;

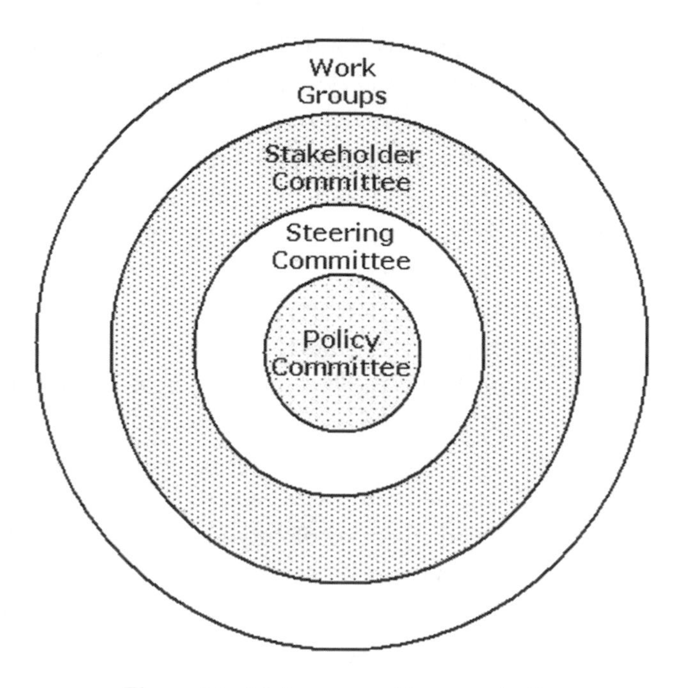

Figure 4.1 Advisory committee organization.

- A task force comprised of supervisory-level operations officers who provide analysis and staff support; and
- Specialist groups that perform technical staff work within their area of expertise.

It is clear that the FIRESCOPE model offers a number of advantages. Planning is actually accomplished by the people who will use the plan. Since membership is open to all organizations with responsibilities under the plan, there is a synergy that takes place as expert knowledge and specialized planning resources are made available. Like the previous model, it allows work to be accomplished by those that are technically competent, allows for broad oversight, and retains the capacity for effective decision making. The FIRESCOPE model also meshes very well with the requirements of NFPA 1600 and EMAP. Figure 4.2 is an example of how one jurisdiction adapted the FIRESCOPE planning model for its emergency management program.

No matter what model is used, the advisory committee is a critical resource in gaining commitment to the emergency management program. It provides for both program oversight and program resources. It allows broad-based community involvement in the emergency management program. However, as will be discussed in another chapter, one must always remember that the advisory committee is used for emergency planning and may not necessarily be operative in managing emergencies. It is a strategic construct and not a tactical one.

Figure 4.2 FIRESCOPE planning model adapted to local jurisdiction.

PROGRAM ADMINISTRATION

Administrative and Strategic Plans

NFPA 1600 and the EMAP Standard require that an emergency management program be institutionalized and documented. While it is possible to demonstrate institutionalization through a variety of separate documents, the most effective way to do this is through an administrative plan for the program. An administrative plan establishes the policies and procedures for managing internal processes. In other words, it lays out the basics of the program: its purpose, concepts, and responsibilities. In the absence of an administrative plan, it is difficult to demonstrate the existence of a comprehensive emergency management program as opposed to a mere collection of projects and initiatives.

In the context of the emergency management program, the administrative plan is an extremely powerful tool. If it is written solely for the local emergency management office, it is relatively meaningless. If, however, it is an organizational document supported by senior management, it can be used to garner support from other organizational elements. It provides for the assignment of responsibility for specific program elements to other parts of the organization and fosters agreement on common goals for the program. More importantly, it provides continuity through the frequent changes in leadership common in most organizations.

Where the administrative plan establishes the operating framework and overall goals for the emergency management program, the multi-year strategic plan provides a long-range strategy for achieving program goals. Where the administrative plan assigns functional responsibility for program elements, the strategic plan details how this responsibility will be fulfilled. While there are no standard formats for these plans, a useful way of organizing these documents is by the stability of the content. Generally, the administrative plan elements tend to stay fairly stable for several years while elements of the strategic plan are reviewed and updated annually.

The administrative plan will generally include:

- Vision and mission statements
- Long range program goals

- Explanation of how the program vision, mission, and goals relate to those of the parent community (i.e. value added)
- Summary of relevant laws and authorities
- Operating structure
- Program policies and procedures
- Sources of funding and normal budget levels
- Provisions for program evaluation

The strategic plan normally covers a period of three to five years, with provisions for annual review and update (an annual review is suggested by NFPA 1600). If the vision changes, the strategic planning process may need to be implemented to develop a new strategic plan. The strategic plan includes the following components:

- Vision and mission statements
- Multi-year program goals
- Multi-year program objectives
- Annual program goals
- Annual program objectives
- Budget forecasts

Developing the strategic plan is a complex process that involves many different aspects. This chapter will discuss the components of the plan and suggest some baseline data that should be considered in developing the plan. The next two chapters will discuss hazard analysis and the development of strategy, two elements that form the core of the strategic plan.

Executive Policy and Strategic Planning

As part of program administration, NFPA 1600 and EMAP require the documentation of executive policy to include vision and mission statements. These policies define the intent of the program and establish the program as a legitimate part of the organization. The formulation of executive policy is in fact the culmination of a strategic planning process that asks questions such as "What is our purpose?" and "Whom do we serve?" This strategic planning process is often overlooked and terms such as "vision statement," "mission statement," and "goals and objectives" are frequently confused or misused. When properly accomplished, strategic planning affords an opportunity for the emergency manager to build a solid foundation for the emergency management program.

The question "What is our purpose" is not as simple as it sounds. It is too simplistic to merely parrot, as most programs do, the phrase "to save lives and to protect property and the environment." This question really forces one to confront the issue discussed in the previous chapter, "What value does the emergency management program add to our community?" If this value is limited to protection only, it tends to have limited resonance with senior managers and politicians. If, on the other hand, the program provides value through the strategic reduction of risk or through efficiencies in the capacity to deal with risk, then the program is taken more seriously. The purpose of the emergency management program must be carefully thought out and clearly phrased to demonstrate the strategic importance of the program to the achievement of the parent community's goals.

The question of "Whom do we serve?" is also problematic. Ultimately, it is the citizens of a jurisdiction who will benefit from an effective emergency management program, but does the program really serve those ultimate beneficiaries directly? The emergency management program most directly serves the community itself as an entity. That is, the program is intended to help manage risk by coordinating the efforts of those subgroups with responsibility for program elements to produce a synergistic reduction in risk or an increase in the capacity to respond. This implies that the true customers of the emergency management program are the organizational entities that it serves: elected officials, department heads, and senior executives.

As the emergency manager conducts an assessment of the program, the answers to these and similar questions become critical. Accurately identifying core functions and clients means that the remaining components of the strategic planning process (i.e. goals and objectives) become more focused and lead to a more efficient program and better use of limited resources. Peter Drucker illustrates this point in his classic work, *Management: Tasks, Responsibilities, and Practices*, using a case study from AT&T's early days.

Based on analysis of AT&T's business from 1905 to 1915, Theodore Vail answered the question "What is our business?" with the response "Our business is service." Vail had grasped several issues that went beyond just the provision of a telephone system. He noted that a natural monopoly such as AT&T was subject to nationalization and that only strong public support could counter this. By correctly identifying the public perception of value-added as service and not the telephone system itself, Vail set the strategic vision for AT&T for the next 50 years. It is also interesting to note that Vail was fired from AT&T in the late 1890s for asking that same question, "What is our business?" He was rehired 10 years later when the government was considering nationalizing AT&T, largely because it was floundering due to a lack of a unifying vision.

The importance of this internal and external analysis to the rest of the process cannot be overstated. This becomes obvious as one considers the rest of the strategic planning process shown in Figure 4.3:

- The vision statement is developed through internal and external analysis and answers the question "Where do we want to be?" Executing the organization's mission fulfills this vision.
- A mission statement that answers the question "What do we do?" is derived from the vision statement. This mission is executed by implementing a strategy.
- The strategy is a long-term plan for executing the mission that answers the question "How do we get from where we are to where we want to be?" The strategy is implemented by achieving strategic goals.
- Goals are broad guidelines used to assist in long range planning and in measuring progress. They answer the question "How do we measure our progress?" and are achieved by meeting objectives.
- Objectives are the key results that must be met to achieve goals. Key results are produced by action. This answers the question "How do we get it done?"
- Action is the end result of the strategic planning process. It is required to accomplish objectives and ultimately leads to the achievement of the organization's vision.

Figure 4.3 Strategic planning model.

A way of understanding this process is to consider an analogy with the military planning system. While the analogy is not exact, it is close enough to understand the dynamics at work. In preparing for battle, a commander articulates a concept for the campaign (a vision). Major subordinate commands interpret their responsibilities under that concept (mission) and develop a plan (strategy) to achieve it. Within that plan are major tasks that must be accomplished and, ultimately, specific assignments to subordinate units (goals and objectives).

Understanding the strategic planning model is critical to overall program management. Each link in the model is co-dependent and must be understood in context. This context is crucial and a single element cannot be divorced from the model and considered separately.

CASE STUDY: OPERATION MARKET GARDEN, SEPTEMBER 1944—THE STRATEGIC PLANNING PROCESS

The planning related to Operation Market Garden in World War II presents a good analogy for the strategic planning process. Note how each element within the process works to achieve the previous element and how all elements ultimately relate to the strategic vision.

Operation Market Garden was a bold plan to break the stalemate along the northern European front and bring World War II to a quick close by striking at Germany's industrial heart. The plan was ultimately unsuccessful in achieving its goals because of an intelligence failure to detect the 2nd SS Panzer Corps, which had been sent to Arnhem to rest and refit.

The Vision:
Destroy German armed forces in the field and force the surrender of Germany.

The Mission:
Seize a crossing across the Rhine River and invade the Ruhr, Germany's key industrial region.

The Strategy:
Use two separate but coordinated assaults to seize a bridgehead across the Rhine. Use airborne troops to seize intermediate bridges and a final bridge across the Rhine, and then follow up with a ground assault to consolidate territory gained and position resources for an assault across the Rhine.

The Goals:
Capture and hold strategic bridges along the route.
Break through the German front line.
Capture and hold a bridge across the Rhine River.

The Objectives:
Eindhoven bridges—U.S. 101st Airborne Division
Nijmegen bridges—U.S. 82nd Airborne Division
Arnhem Bridge—British 1st Airborne Division
Ground Assault—British XXX Corps

Many organizations spend countless hours arguing over every syllable in a vision statement without grasping that it is not the statement that is important but rather the entire strategic planning process that ultimately defines organizational success.

Figure 4.4 provides an example of a vision and a mission statement from California's FIRESCOPE program. Note that the vision statement is a brief and clear statement that encapsulates the organization's definition of success while the mission statement explains succinctly the steps that are being taken to achieve the organization's vision. Compare this with the mission and vision statement of a public safety dispatch agency in a large city shown in Figure 4.5. Note the very broad scope of the vision compared to the very narrowly defined mission. There is an obvious disconnect between the organization's desire to be an international model and its mission of linking customers and

Vision

The FIRESCOPE Vision is to lead in the development and enhancement of Fire Service partnerships in California and nationwide by promoting the use of common all risk management systems for planned and unplanned events through the innovative use of technology.

Mission Statement

The Mission of FIRESCOPE is to (1) provide professional recommendations and technical assistance to the Director of OES (Office of Emergency Services) and the OES Fire and Rescue Branch on the following program elements:

- Statewide Fire and Rescue Cooperative Agreement (Mutual Aid) Plan.
- Statewide Fire and Rescue Cooperative Agreement (Mutual Aid) System.
- Mutual Aid Use and Application.
- OES Fire and Rescue Branch staffing needs
- Policies and Programs.
- Apparatus and Equipment programs.

And (2) maintain a system known as the FIRESCOPE "Decision Process" to continue statewide operation, development, and maintenance of the following FIRESCOPE developed Incident Command System (ICS) and Multi-Agency Coordination System (MACS) components:

- Improved methods for coordinating multi-agency firefighting resources during major incidents.
- Improved methods for forecasting fire behavior and assessing fire, weather and terrain conditions on an incident
- Standard terminology for improving incident management.
- Improved multi-agency training on FIRESCOPE developed components and products.
- Common mapping systems.
- Improved incident information management.
- Regional operational coordination centers for regional multi-agency coordination.

Figure 4.4 FIRESCOPE vision and mission statements.

Vision Statement

The vision of the XXXX Communications Department is to be the international model recognized for leadership, professionalism and innovation in 911 public safety communications.

Mission Statement

We, the members of the XXXXX Communications Department are committed to excellence in public safety, providing the linkage between the residents and visitors of XXXXX 's diverse city and its emergency services resources. We take pride in dedicating ourselves to professionalism and public service.

Figure 4.5 Public Safety Dispatch Agency vision and mission statements.

emergency services. Rather than serving the vision, the mission is more a statement of organizational values. Execution of this mission will not necessarily result in achievement of the vision. Further, it is unclear how this vision will add value to the community it serves. Vision and mission statements are not just lofty sounding statements but must be carefully linked through the strategic planning process.

It is also worth noting that visions and missions can change over time. Generally, a vision is fairly stable, with a life expectancy of anywhere from five to 30 years. The AT&T vision noted earlier in the chapter worked for 50 years. Mission statements are also fairly stable, usually lasting between three to five years. Multi-year strategies are generally written for five-year periods but are usually adjusted annually to account for changing goals and objectives.

Developing the Strategic Plan

The strategic plan is the means by which the organization's strategy gets implemented through identifying goals and objectives and fixing responsibilities for achieving them. The content of the plan is based on a process that considers the community's risk, the strategies it will use to manage risk, and the resources needed to implement these strategies. These subjects will be covered in succeeding chapters. This chapter will consider the mechanics of the plan and its relationship to the emergency management program.

The strategic plan establishes the long-term goals that the community wishes to achieve and the process by which this will be accomplished. In other words, the plan establishes what the community expects to accomplish over a specific time period and who has the responsibility for making this happen. In its simplest form, it is a collection of strategic goals and objectives. However, the strategic planning process is the glue that binds these elements together. A strategic plan is not just a collection of disparate goals but the means by which these goals are focused to meet the community's definition of success. Each strategic goal must contribute to mission accomplishment and ultimately to the achievement of the community's vision.

The terms "goals" and "objectives" are sometimes confusing in that the terms are essentially synonymous. In fact, many management texts refer to general objectives and performance objectives rather than to goals and objectives. Goals or general objectives are broad guidelines to assist in planning. Objectives or performance objectives are specific and measurable. In other words, a goal is a broad or strategic statement of what the community wants to accomplish while the objectives explain how that goal will be achieved tactically. Objectives can be further broken down into specific work assignments or tasks.

Because they are so broad, defining goals is generally easy for most communities, particularly if the appropriate work has been done on the vision and mission statements. However, developing performance objectives involves a realistic appraisal of available resources and a willingness to resolve conflicts through prioritization and compromise. The National Management Association has developed a checklist for evaluating performance objectives (Figure 4.6) that provides a simple methodology for establishing useful objectives.

The development of performance objectives presents an opportunity for defining roles and responsibilities for various elements of the program. This is the chance for the emergency manager to develop support for program goals and to foster involvement and support from organizational elements. This process requires agreement from organizational elements that they will 1) accept lead responsibility for a specific program task and 2) commit appropriate staff resources to accomplish assigned objectives.

Because a strategic plan is a multi-year plan, objectives can be long term, with supporting tasks forming part of an annual work plan. This

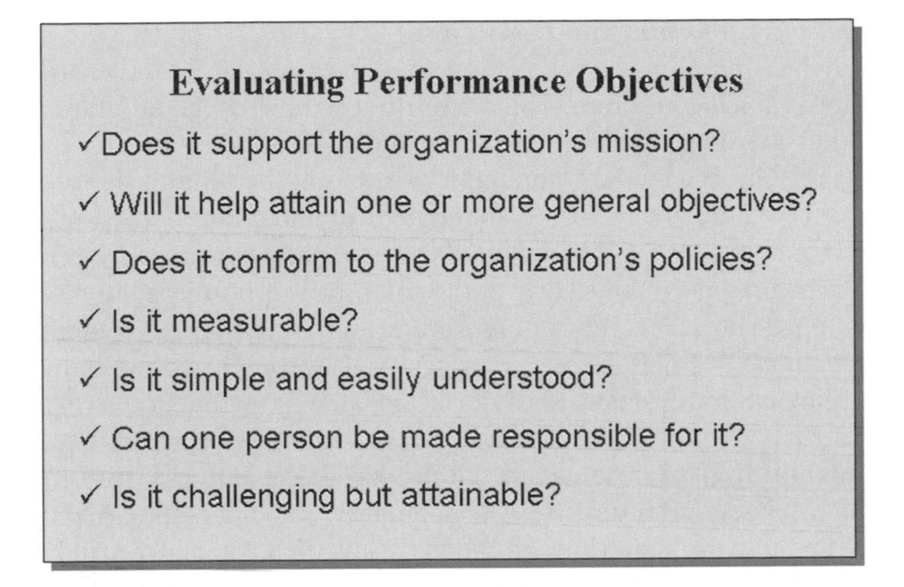

Figure 4.6 Checklist for evaluating performance objectives, *The NMA Handbook for Managers,* National Management Association, 1987.

allows for long-range planning and incorporation of projects into lengthy budget cycles, as well as providing for program continuity during changes of administration.

As one goes through the strategic planning process, an important starting point is the question "What should we be doing?" This question is ultimately at the heart of the strategic planning process. It implies an identification of potential problems and produces the strategy that will be used to solve them. To answer this question requires a two-part process. In the first part, one must identify what it is that the community is mandated to do by law. The second part of the process builds on this base to actually determine what "should" be done based on public expectation and best practices from a variety of disciplines.

Enabling Authorities and Legislation

A key starting point in developing the strategic plan is the enabling authority for the program. This authority establishes the program and outlines the duties and authorities of the program coordinator. From this legislation or policy statement one should be able to derive the specifics of the emergency management organization's mission and have some idea of the intent behind the law, policy, or regulation.

Unfortunately, many such authorities in the public sector were developed during the civil defense era and are extremely outdated. Such authorities may be vague or scattered through various statutes. Enabling authorities have rarely been updated to include modern concepts such as hazard analysis, mitigation, and recovery planning and frequently limit themselves to the preparedness and response functions of classic civil defense. Consequently, the emergency manager may find that the authorities provided by law do not actually match the duties that have accreted to the office over the years. In some cases, this could severely limit the scope of the program or define authorities so broadly that the emergency manager is overwhelmed by duties that cannot be delegated to other agencies.

In developing overall program strategy, therefore, it may be necessary to perform a legislative update of the enabling authority for public sector agencies or to use it simply as a starting point and not allow it to limit the scope of the program. It is also important to realize that legislation must be updated periodically. In addition, taking the strategic view requires that the emergency management program be able to comment on, support, initiate, or oppose emergency management related legislation with potential impact on the program or jurisdiction. Consequently, NFPA 1600 and EMAP require a documented strategy addressing this need for legislative and regulatory revisions.

There are also tasks that the community may be required to accomplish by law. As one reviews these requirements, it is well to keep in mind the hierarchy of laws and the authority from which a requirement is derived. Statutory requirements begin with the passage of a law by a legislative body. This law may be given a name based on the purpose or the originators and may have both a long name and a short name. The law is then incorporated into the legislative entity's statutes. In some cases, the law has sufficient detail for implementation, but in many cases, a series of regulations is developed to implement the provisions of the act. Based on these regulations or codes, the agency charged with implementing the act may develop procedures and guidelines for accomplishing the purposes of the act. It is important to note that the act and the regulations have the force of law but, in most cases, there is some latitude for negotiation with agency regulations and guidelines.

Complicating things even further is the authority of an elected official to issue executive orders to quickly implement new policies. For example, many Homeland Security offices across the nation have been established through executive order rather than through legislation, leading to confusion over their roles and authorities in relation to organizations established by statute.

For the public sector, there are several key authorities that form the basis of the emergency management program. Some of the main ones are:

- The Robert T. Stafford Disaster Relief and Emergency Assistance Act—the Stafford Act is the legislative descendent of the Civil Defense Act of 1950 and the Federal Disaster Act of 1950. It includes the federal government's authorities for disaster relief and recovery and establishes grant programs for mitigation and preparedness.
- Disaster Mitigation Act of 2000—DMA 2000 is an amendment to the Stafford Act that requires the development of a hazard mitigation plan by states as a prerequisite for receiving post-disaster mitigation funds.
- 44 Code of Federal Regulations—44 CFR contains the implementing instructions for the Stafford Act. It defines FEMA's policies, restrictions, and procedures for administering grants and programs.
- Superfund Amendments and Reauthorization Act (SARA) Title III—SARA Title III is a 1986 amendment to the Comprehensive Environmental Response, Compensation, and Liability Act (CERCLA) of 1980. It establishes requirements for federal, state and local governments, Indian tribes, and industry regarding emergency planning and "Community Right-to-Know" reporting on hazardous and toxic chemicals.

- 40 Code of Federal Regulations—40 CFR deals with environmental protection. It contains implementing regulations for SARA Title III.
- Homeland Security Presidential Directive 5—HSPD 5 identifies steps for improved coordination in response to incidents. It requires the Department of Homeland Security to coordinate with other federal departments and agencies and state, local, and tribal governments to establish a National Response Plan (NRP) and a National Incident Management System (NIMS).
- Homeland Security Presidential Directive 8—HSPD 8 describes the way federal departments and agencies will prepare. It requires DHS to coordinate with other federal departments and agencies and state, local, and tribal governments to develop a National Preparedness Goal.

Added to this federal legislation are numerous state and local laws and regulations. Much of this legislation is out of synch with current practice and may reference archaic laws or organizations that no longer exist. Frequently, these references find their way into program documents such as the emergency operations plan, resulting in the potential for confusion during times of crisis. It may also create a situation where the community does not have the legislative authority needed to accomplish disaster tasks such as mandatory evacuation or medical quarantine. This situation highlights yet again the importance of a process for revising existing statutes. It also demonstrates why the emergency management program must be broad based: few emergency managers are lawyers; having access to the community's legal and legislative teams is critical.

Grant Requirements

Another source of input for the strategic plan are the performance plans required under a variety of emergency management and homeland security grants. Grant applications require development of a proposal in which the jurisdiction commits to developing certain work products and/or performing certain activities within a specific period of time. Since these work plans will require the use of jurisdictional resources and may, as in the case of the Homeland Security Grant Program, involve multiple agencies, they should be included in the strategic plan. Including grant performance plans in the strategic plan also ensures that these plans are coordinated with the rest of the emergency management program. This results in more effective use of the grant funds by leveraging them to address multiple hazards in addition to terrorism.

The challenge of grants and the reason why their inclusion in the strategic plan is essential is that there is no single source or method of distribution for federal funds. While the Department of Homeland Security has made great strides in this direction, multiple funding streams through various agencies are still the norm. Consequently, in order to maximize use of these grants and coordinate their required activities across departmental lines, it is critical that the jurisdiction have a cohesive strategy that encompasses these grant requirements and identifies potential uses for future grants.

Best Practices

Legislative requirements specify only the bare minimum for an emergency management program. They answer the question, "What must we do?" However, effective emergency management programs must also consider the question, "What should we do?" To round out the program and to ultimately achieve the organization's mission, one must turn to best practices from multiple disciplines and incorporate them into the strategic plan.

"Best practices" are those practices that have become so standard and accepted within a given industry that failure to include them in a program would be considered negligent, whether or not their inclusion is required by law. For example, laws do not mandate many of the components of emergency planning such as the CEM model and the components of an emergency plan. These practices are based on training courses and guidance documents developed by FEMA. A jurisdiction that does not consider the four phases of the CEM model in its planning or failed to have an emergency plan would be considered negligent in any investigation following a disaster. Such negligence could well represent a significant liability exposure for the jurisdiction.

Best practices may also include forecasting future needs based on current industry practices. For example, at the time of this writing there exists no specific requirement for Continuity of Operations Planning (COOP) for local jurisdictions. However, federal agencies have been required to develop COOP plans since 1998 and the concepts extend back to the early 80s. Recent Government Accounting Office reports have catalogued a failure of this effort, resulting in increased emphasis on COOP by congress. In addition, NFPA 1600 and EMAP require continuity plans as a condition of accreditation and the cost of developing these plans are allowable costs under the Homeland Security Grant Program and the Emergency Management Performance Grant Program. COOP planning will no doubt emerge over the next few years as a best practice for local governments and this fact should be considered during strategic planning.

The issue of accreditation is itself another example of this principle. The National Preparedness Standard is voluntary, as is EMAP. However, as was noted earlier, FEMA has funded a baseline EMAP assessment for each state and the FY 2006 EMPG required that states use the EMAP Standard as a baseline in developing their performance plans. It does not take any great leap of logic to infer that conformance with EMAP may one day become a requirement for federal grant funds, making accreditation a potential long-range strategic goal.

The EMAP Standard represents an accepted standard that can be used to impartially assess the emergency management program. The results of an initial baseline assessment can be used to identify shortfalls in the existing program that can then in turn be addressed in the strategic plan. A periodic reassessment can then be used to measure progress. The EMAP Standard is a de facto best practice for the emergency management profession.

The use of best practices has been encouraged in other disciplines for years, usually through the auspices of a national organization. For example, associations like the International Association of Fire Chiefs and the International Association of Chiefs of Police have contributed greatly to defining accepted standards of performance for public safety services. Private sector organizations such as the American Society for Industrial Security, International, and the Disaster Recovery Institute, International have done the same for the private security and business continuity disciplines. Best practices and industry standards offer a wealth of ideas for strategic goals that are achievable and measurable.

Defining Goals and Objectives

The point of looking at indicators such as legislative requirements and best practices is to provide a general direction for the strategic plan. These are factors that must be taken into account as one develops concepts and ideas for community response. The strategic plan is a description of how the various components of the emergency management program will be implemented and is, in effect, the result of many separate strategies. For example, if one's strategy is to coordinate all phases of a response (emergency response, continuity, recovery, etc.) from a single emergency operations center, then the plan might include goals related to capital projects and staff development.

Goals in the strategic plan do not drive the emergency management program but are instead driven by the program. The strategic plan is the outgrowth of the strategic planning process—it is the mechanism by which disparate elements of the program are tied together to ensure that they are working toward a common vision. This allows for coordination by the emergency manager while at the same time allowing decentralized planning by other agencies. Goals reflect what the community has collectively decided to do and begins translating strategy into actual accomplishments.

The real meat in the strategic plan, though, are the performance objectives. The secret to defining objectives is to keep them realistic and measurable. Referring back to Figure 4.6, the National Management Association suggests several key elements that can be used in defining objectives: consistency with vision, goals, and policies, simplicity, measurability, and attainability. To conform to these elements, an objective should have the following components:

- A description of what is to be done in simple terms that makes it easy to determine when the objective is met.
- Assignment of responsibility for achieving the objective. Wherever possible, this should be a single agency or individual so that the responsibility doesn't become diffused. It is better to assign it to the lead agency that chairs a task force, for example, than to the task force itself. The lead agency can be held directly accountable where the task force generally cannot.

- A timeline and milestones to assist in measuring progress. Open-ended objectives are seldom achieved because it is too easy to defer the project. Long term projects that do not have reporting milestones tend to be ignored or deferred in favor of more pressing issues.
- Budgetary data that estimates the resources needed to achieve the objective. This can be hard costs, such as the purchasing of equipment, or soft costs such as internal personnel costs. Without a cost estimate, it is not possible to determine whether the objective is realistic and achievable. To be realistic and achievable means that decision makers understand the resources that will be required to achieve the objective and are willing to provide them.
- Proposed source of funding. If the objective does not include a description of how the costs associated with it will be met, it isn't really achievable. Objectives without funding are too easily deferred on the basis of "there's no money for that." In this age of tight municipal budgets, part of defining an objective is determining how it will be funded.

To be effective, the goals and objectives must ultimately be translated into action through incorporation into the annual work plans and budget submissions of subordinate agencies. Figure 4.7 shows the interrelationship between the strategic plan and other administrative documents. This integration into work plans and budget is the reason that broad-based participation in developing program plans and strong support from senior executives are critical to the success of the program.

STRATEGIC PROGRAM ELEMENTS

The purpose of the emergency management program is to coordinate enterprise-wide planning related to program goals and strategies. As has been noted, the mechanism for this coordination is the strategic plan. While a good portion of this plan will be devoted to preparedness issues, there are three key strategic elements that should be put in place to support planning efforts: resource management, training, and finance. These functions have implications for the strategic plan.

Administrative Plan

"The lead agency for hazard mitigation plan development is the city planning department. Funding for developing and sustaining such plan will be included in the annual department budget."

Strategic Plan

Goal	Objective	Estimated Completion	Lead Agency	Est. Cost	Funding Source
Reduce risk vulnerability through a multi-hazard mitigation program	Assess vulnerability to natural hazards	Year 1	Planning	$50K	Department Budget
	Develop Mitigation plan	Year 2	Planning	$30K	HMGP

Department Budget

Project	Staffing	Admin Costs	Budget Request
Conduct natural hazards vulnerability assessment	, 75 FTE	$15, 375	$52, 739

Figure 4.7 Implementing program plans. Note the relationship between the functional responsibility in the administrative plan, performance objectives in the strategic plan, and the department budget.

Resource Management

An important part of the strategic plan is the identification of resource management objectives. This is not the same as logistics, which will be discussed in a later chapter, but is instead the strategic process by which the jurisdiction addresses potential resource shortfalls. This process begins with an analysis of the resources required to implement the strategies developed for dealing with particular hazards. For example, if the jurisdiction has a strategy that requires sheltering 10,000 citizens, one can estimate the personnel, equipment, training, facilities, and costs associated with meeting this requirement. The jurisdiction next conducts a survey to determine what resources currently exist and identifies shortfalls between what is on hand and what is needed. The jurisdiction then determines a strategy for meeting these shortfalls through stockpiling resources, conducting training, or developing contracts or mutual aid agreements. This strategy is then translated into objectives that are incorporated into the strategic plan.

A second component of resources management is the ability to locate needed resources at the time of crisis. The normal assumption for many programs is that this means that the program coordinator must maintain a master list of all available resources. Given the limited resources available to most emergency management offices, it is unlikely that such a list can be assembled or maintained with any accuracy. However, if one uses the paradigm that the emergency manager is the coordinator of an enterprise-wide program, the issue is no longer how the emergency manager can maintain such a list, but instead a question of how the jurisdiction manages its resources.

Once one grasps that it is the jurisdiction as a whole that must manage resources, it becomes acceptable for departments and agencies to maintain their own resource inventories and it becomes the responsibility of the emergency manager to determine how these inventories can be made accessible during a crisis. This makes sense when one considers how resources are managed on a daily basis. Heavy equipment, such as cranes and bulldozers, is expensive and does not sit idle. If they are private sector resources, they may move in and out of the jurisdiction in response to customer requirements. Maintaining a list of type and location at the emergency management office would be labor intensive to the point of impossibility. However, the owners of that equipment know exactly where each item is at all times. The focus of the emergency management program should not be on the development of resource inventories but on the process of accessing existing inventories.

This highlights one of the advantages of viewing the emergency manager as coordinator rather than executive agent. It allows the use of existing systems and resources to accomplish program objectives. There is no need to invent a new system for logistics for use in disasters. One examines the existing organizational system, identifies issues that would hinder operational response, and develops strategies to remove those roadblocks.

A potential source for filling resource shortfalls is through the use of mutual aid agreements. Mutual aid resources are made available from other jurisdictions under pre-existing agreements. These agreements resolve issues that can hinder response such as insurance coverage, payment for services, etc. While mutual aid is usually given without charge, many jurisdictions are rethinking this policy to allow for reimbursement under federal disaster assistance programs. For example, some jurisdictions will provide resources at cost for the first 12 hours, then bill the receiving jurisdiction for the full cost of services after the 12-hour trigger is reached.

Unfortunately, most jurisdictions view mutual aid as primarily law enforcement or firefighting resources because these are the resources most likely to be used on a regular basis. Few states have adopted a master mutual aid agreement that allows for the free exchange of any resource among signatories to the agreement, such as social service workers or public works staff. Even fewer have begun a process to manage these types of mutual aid resources through a coordinated process. This means that for much of the country, there are limited authorities for sharing resources and no formal mechanisms to locate and dispatch them.

Mutual aid occurs at the state level under the Emergency Management Assistance Compact. While there have been concerns raised that EMAC is administered through an independent organization and not through the mechanism of the National Response Plan, there is evidence from Hurricanes Katrina and Rita that suggests that the system works well. As of this writing, EMAC is a system still very much under development and has been integrated into only a few state plans.

The concept of a national mutual aid system leads to the need for precision in requesting resources. This is a lesson that FIRESCOPE planners learned the hard way and it led to a system of resource typing for fire equipment. To understand the importance of resource typing, consider the situation faced by fire fighters in a major wildland fire. There is clearly a need for additional apparatus, so the chief on scene asks for a fire engine. However, fire engines come in various sizes. They can also be deployed with a crew or without a crew. In some jurisdictions a crew consists of five fire fighters, in others it may be four. How does the chief ensure that he or she gets what is needed? Resource typing allows the chief to formulate a precise request for the resources required.

Resource typing is, as one might expect, a complex task. FEMA is currently engaged in a major project to type the various types of resources need in disaster response. Putting it in perspective, however, resource typing is important for requesting discipline-specific personnel or equipment. It is not so critical for other operational requirements. There are times when being too specific can have a negative impact on response. On one FEMA operation in the Pacific, an inexperienced deputy regional director specified a particular type of aircraft to be used for delivery of relief supplies. The delivery was delayed several days while the specific type of aircraft was located and made available by the U.S. Air Force. It was later discovered that the Coast Guard had two smaller aircraft that could have made the delivery immediately. By specifying the method of delivery rather than the requirement for cargo capacity, the deputy regional director limited the options of the transportation section and delayed the operation.

Training

When one considers training under the emergency management program, it is important to remember that this does not refer to all training that is done by the jurisdiction but rather the training needed to support the program. Like the resource inventory, this means that the emergency manager does not need to track all emergency management-related training directly but should have a process to access existing records as needed and for developing specific training related to plans and strategies, such as training on incident management software.

The identification of training objectives begins with a needs assessment. A good starting point is to consider any regulatory requirement. For example, the Homeland Security Grant program has specific requirements related to training on the National Incident Management System. Acquisition of protective equipment, such as gasmasks or HAZMAT suits, triggers OSHA training requirements. Mandated training may also include the requirement for periodic refresher training or recertification.

The real core of the assessment is an examination of plans and strategies and a determination of the skills required to implement them. The scope of such training is related to the role the individual will play in response to crisis. The curricula for the Incident Command

ICS Curricula

ICS-100: Introduction to ICS

ICS-200: Basic - ICS for Single Resources and Initial Action Incidents

- •Principles and Features of ICS
- •Organization Overview
- •Incident Facilities
- •Incident Resources
- •Common Responsibilities Associated with ICS Assignments

ICS-300: Intermediate - ICS for Expanding Incidents

- •Organization and Staffing
- •Organizing for Incidents or Events
- •Incident Resources Management
- •Air Operations
- •Incident and Event Planning

ICS-400: Advanced - ICS Command and General Staff—Complex Incidents

- •Command and General Staff
- •Unified Command
- •Major Incident Management
- •Area Command

ICS 401 – Multi-Agency Coordination

ICS 402 –ICS for Executives

Figure 4.8 ICS Curricula—National Wildfire Coordinating Group.

System are good examples of the relationship of scope to responsibility. Courses range from an entry-level awareness course that can be taken by all personnel to a course specific to those with high-level responsibilities in an incident.

The purpose of the needs assessment is to identify the type, scope, frequency, and target audience of training needed to support the emergency management program. However, this does not mean that the program coordinator has responsibility for conducting the identified training. In most jurisdictions, there are internal resources available. Departments routinely conduct training and maintain appropriate records. The responsibility of the emergency manager is to develop a process for leveraging these existing resources to support the program.

This allows for as much training as possible to be institutionalized within the existing system and allows the emergency manager to focus on the gap between existing resources and identified need.

In closing this gap, the emergency manager is not so concerned with the training itself as he or she is with institutionalizing it. This means that instead of teaching the class directly, the emergency manager should consider alternatives like contracting for a train-the-trainer course to build a cadre of instructors or consider a partnership or contract with the local community college. Options such as these provide a long-term benefit and help to normalize training within the community.

The same is true of training administration. Instead of developing a system for maintaining records, the emergency manager should consider: 1) What information do I need to have access to in order to ensure training has been done, and 2) Who is already maintaining or could maintain this information? It is much easier to compile summary data than it is to maintain individual records.

The process of identifying and fulfilling training needs should lead to a series of objectives that identify what training will be given, who will be involved, and who will provide the instruction. This process should also provide for an estimate of costs and the identification of potential funding sources. These objectives are then incorporated in the strategic plan to coordinate decentralized training.

Finance

Government is very practiced at developing initiatives that are essentially unfunded mandates. This means that the initiative is held up as a lofty goal but those charged with implementing the initiative must do so within the constraints of normal budget and staffing levels. In actual practice, this means that such plans are limited in their implementation to those things the jurisdiction would have accomplished anyway or to those things that it is forced to do by public or legislative pressure. This is particularly true in the emergency management discipline where program budgets are usually small and many of the major tasks to be accomplished must be funded through separate agency operating budgets. If these agencies are not committed to supporting the emergency management program, these tasks may not be accomplished owing to competition for funding with the agencies' internal priorities.

An important part of strategic planning is determining how the entity will actually pay for components of its emergency management program. In general, tasks without an identified budget source will not be accomplished. These funds may be from an external source such as a federal grant or corporate donation or may come from a commitment to use existing or redirected funds from an internal budget. Budgeting within the context of the strategic plan offers a number of advantages:

- It documents the commitment of a department to use internal funds in a particular budget cycle for a specific task.
- It provides a strategy for the use of present and future grants.
- It identifies shortfalls in future budget projections, providing time to identify alternative funding.
- It documents actual levels of funding for the entire emergency management program instead of just for the emergency management office, demonstrating the jurisdiction's commitment to the program.

The budget component of the strategic plan is by nature very broad and represents an estimate rather than actual costs. Actual costs would be determined as part of the normal budget development cycle. It is also subject to adjustment as goals and objectives are modified or as priorities change. Nevertheless, adding a budget component to performance objectives goes a long way to ensuring that the objectives are in fact achievable and provide long-range continuity for the emergency management program.

Budgeting for program support is not the only financial challenge for the emergency management program. Planners must also consider what emergency procedures must be in place to respond to crisis. These procedures must address two main concerns. The first is how to streamline the purchase of goods and services. Most jurisdictions use a complex system for contracting to prevent fraud and ensure all procedural requirements are met. In a crisis, goods and services must be purchased immediately and normal contracting requirements are usually waived when emergency authorities are invoked. However, if the financial services staff are not familiar with these emergency authorities and have not planned for their use, considerable delays can ensue.

During Hurricane Katrina, the establishment of a base camp to support mutual aid firefighters was delayed because it took several weeks to find an official from New Orleans who could authorize the costs, even though FEMA had committed to reimburse the cost.

A second major concern is the tracking of costs associated with the response. Many of these costs are recoverable through federal relief programs and/or insurance, but having acceptable documentation is critical. Establishing appropriate accounts and tracking codes should be done before a crisis so that costs can begin to be accumulated immediately.

The resolution of these financial issues would be difficult for the emergency manager acting alone. However, in an enterprise approach, financial services staff can be brought into the process. As planners identify needed authorities or processes, the financial services staff can identify what changes to authorities would be required to meet these needs or can suggest other approaches if necessary.

PROGRAM EVALUATION

Program evaluation is an ongoing process that involves a regular review of the accomplishment of strategic plan objectives, maintenance of program elements against an established standard, and a formal corrective action program to address deficiencies identified in program reviews and tactical exercises and operations. This program should be formally established, centrally administered, and documented in the administrative plan.

With any program, sooner or later one must ask the question, "How well are we doing?" In more prosaic terms, this question could also be asked, "What are we getting for all the money we are spending on this program?" Like anything else in emergency management, the question of program effectiveness can be measured on several levels.

Quantitative Assessment Tools

Quantitative assessment tools are those that measure the program against specific criteria such as pre-identified metrics or a standard such as EMAP. They usually are fairly objective (either you met the required criteria or you didn't). They lend themselves well to a checklist format, making them fairly easy to use. Such tools are useful in determining conformance with standards or compliance with legislative requirements and can generate easily understood summary reports.

The strategic plan represents an important quantitative tool for evaluating the program. The objectives in the strategic plan are performance measures that can demonstrate progress toward achieving the strategic goals agreed to by the community. If crafted properly, each objective has a metric that determines when the objective has been achieved (e.g. a product developed, training delivered, equipment purchased, etc.), making it easy to determine if the objective has been achieved, in whole or in part, or whether it has not been achieved. Part of the annual review of the strategy should be a summary for senior management of how well annual goals were met. This allows the emergency program manager to present concrete evidence to oversight groups that progress is being made toward achieving the organizational vision and to justify costs associated with the program.

Another important tool for assessing the emergency management program is the EMAP Standard. As has been noted, a baseline assessment against the standard can be an important part of developing the strategic plan. Periodic reassessments against the standard can demonstrate progress or identify new problem areas. Such a reassessment program has an added advantage. EMAP accreditation is valid for a period of five years after which the jurisdiction must seek reaccredidation. If documentation is not maintained during the accreditation period, reaccredidation is extremely labor intensive. It is therefore in the best interests of an entity to establish a program to periodically verify conformance with the EMAP Standard. The results can be incorporated into the corrective action program or the strategic plan as appropriate.

Although they are frequently deconstructed into checklists, the EMAP Standard and its source document, NFPA 1600, are management documents and do not lend themselves well to a checklist format. This is because many of the elements of the standards incorporate a considerable amount of material in a short section and assessing against that element requires the review of multiple documents. As long as this complexity is understood, then a checklist that summarizes findings such as the Local Capability Assessment for Readiness

(LCAR) tool developed by FEMA or commercially available audit tools can be useful.

Quantitative assessments are particularly useful at the operational level, where tasks are more easily defined. This is the theory behind the development of the DHS Target Capabilities List. This list identifies 36 capabilities that DHS considers essential for performing the critical tasks identified in the Universal Task List. The expectation is that these capabilities should be developed and maintained, in varying degrees, by each level of government. The capabilities include performance measures, making the Target Capabilities List a potential tool for assessing performance.

As of this writing, DHS is refining the Target Capabilities List to improve its utility as a measurement of performance by including more precise metrics. In a paper titled "Measuring Prevention," Glen Woodbury, a graduate of the Naval Postgraduate School's Homeland Security Program, suggests one way in which this might be accomplished. Woodbury suggests a simple scale that measures program outputs and provides a methodology for assessing progress in achieving desired outcomes. The simplicity of the Woodbury Scale makes it exceedingly adaptable to a variety of program elements.

Qualitative Assessment Tools

Qualitative tools are subjective and attempt to assess the emergency management program through actual performance. Where a quantitative tool might ask, "Do you have a plan and does it address functions 1-11?" the qualitative tool seeks to answer the question, "How well does your plan work?" Evaluating the program from a qualitative perspective is difficult but gives a better indication of the program's capabilities than does a quantitative assessment. The two most common qualitative tools used by emergency managers are the exercise and the after-action reports of actual incidents.

Exercise Programs

A major measure of an emergency management program's capacity to respond to crisis is the performance of program elements under the stress of a simulated crisis. Exercises offer a number of benefits to the emergency management program for:

Woodbury Scale

0 = No effort or system underway or no recognition of the need

1 = Recognition of the need but no effort or resources to accomplish the output

2 = Initial efforts and resources underway to achieve the output

3 = Moderate progress towards accomplishing the output

4 = Sustained efforts underway and output near to fulfillment

5 = Output achieved and resources devoted to sustain effort

Figure 4.9 Woodbury Scale, "Measuring Prevention," Glen Woodbury 2005.

- Training in roles and responsibilities in a risk-free environment
- Assessing and improving performance
- Demonstrating the resolve of the jurisdiction to prepare for crisis
- Improving interagency coordination and communications
- Identifying resource shortfalls.

To be effective, however, exercises must be part of a comprehensive exercise plan that over time tests all components of the emergency management program. This means that exercises must be considered in multi-year cycles and not as individual events. Ideally, exercises build on each other by increasing in complexity, scope, and scale. The exercise plan, by virtue of its multi-year cycle, is part of the strategic planning process and objectives related to the program form part of the strategic plan.

In addition to considering exercises as separate activities, another common problem is the identification of the exercise with a hazard. For example, a jurisdiction will begin planning for an exercise by deciding that the exercise will be based on terrorism or an earthquake.

The problem with such an approach is that it focuses the exercise on the hazard rather than on evaluation of the emergency management program. In the initial stages of exercise planning, the hazard that triggers the exercise is irrelevant. The preferred approach is to first identify exercise objectives that are consistent with the multi-year exercise program and the strategic goals of the community.

Exercise objectives are the true focus of the exercise. They define what it is the exercise is supposed to accomplish and provide a metric against which the success of the exercise can be measured. Examples of such objectives might be to test the performance of new communications systems or to identify revisions for emergency operations center procedures. It is also not uncommon for subordinate agencies to develop their own internal objectives, such as testing the interface between the EOC and the department operations center. However, overall exercise objectives should be limited to a small number to keep the scope manageable.

Once exercise objectives are established, it is then appropriate to determine the scenario for the exercise. The exercise planners then have the option of selecting a suitable scenario, or, if a particular scenario is required (for example, by a grant requirement), that scenario can be tailored to produce the conditions needed to test the objectives. While there are qualitative and quantitative differences among scenarios, for many objectives these may be incidental. For example, if the objective is to test evacuation plans, one may select a terrorist chemical attack, a large HAZMAT release, a hurricane, or a conflagration as the initiating agent. While variables such as time to onset or overall population risk may vary, the requirement to evacuate does not change.

Exercise design is best accomplished by creating a team with representations from the agencies that will be participating. This ensures that issues relevant to those organizations are considered in the planning. A product of the exercise design team is the exercise plan, a document that contains the administrative information relative to the exercise: objectives, timeframes, work assignments, location, etc. The plan should also include copies of any other materials produced such as player guides, facilitator manuals, etc. Many exercises do not go well because these administrative details have not been planned and documented.

There are considerable resources available for developing exercises. Among the best of these are the four volumes of the Homeland Security Exercise and Evaluation Program and the FEMA exercise design course. The FEMA course is available both as an instructor-based course or online through the Emergency Management Institute.

Exercises are divided into two broad classes: discussion-based and operations-based. Discussion-based exercises are those that focus on policy-based issues. They provide for guided discussions among participants without the stress of actual operations. They are excellent for focusing on processes and interagency issues and make good starting points for the multi-year exercise program.

The most common discussion-based exercise is the tabletop exercise. Tabletops are extremely versatile and can be tailored to almost any timeframe or budget. A tabletop exercise focusing on a single objective can be done in as little as 10 minutes. A complex multi-agency tabletop can take several days. One emergency manager in California has even devised a five-minute tabletop for use at the opening of regular jurisdictional staff meetings. The key to successful tabletop exercises is to remember that they are guided discussions. This means they must have clearly defined exercise objectives, make use of a facilitator, and be built around specific questions. It is not sufficient to provide a scenario and say, "What do you do?" Instead, each part of the exercise should be designed to elicit discussion on issues relevant to the exercise objectives.

Discussion-based exercises are excellent tools whose frequent use can help identify poor planning assumptions and eliminate misunderstandings. However, the true measure of response capacity is the operations-based exercise. Operations-based exercises tend to focus on tactical and operational issues by placing the participants under stress. The exercise attempts to approximate as closely as possible the conditions under which the particular system being tested must function.

Operations-based exercises consist of three levels:

- Drills focus on a single specific function or operation, usually performed at the tactical level. An example might be an exercise that tests the ability of the police bomb disposal squad to work in Level A hazard suits.

- Functional exercises test multiple functions or interdependent functions. These are normally conducted at the functional level and focus on coordination issues. An example would be an EOC exercise that tests the interface between the EOC and department operations centers during a simulated hurricane. All elements outside the EOC or the player location are simulated. Functional exercises require a moderate level of resources and around six months of advance planning.
- Full-scale exercises are complex operations and designed to test many different functions in real-time. Field resources are deployed and multiple levels of response are activated. Full-scale exercises require a considerable commitment of resources and require at least a year of advance planning. However, they are the best method for testing a jurisdiction's emergency management capacity.

The success of an exercise ultimately rests on the evaluation process that is used. This process should capture issues in two areas: conduct of the exercise and performance. The first area relates to the administrative component of the exercise such as how well the exercise met its objectives. This information is used to improve the design of future exercises. Performance issues pertain to problems observed in performing tasks or functions or issues with plans and policies. This information is used to make changes within the emergency management program and may have an impact on the strategic plan.

The *Homeland Security Exercise and Evaluation Program Volume 2: Exercise Evaluation and Improvement* provides an eight-step methodology for assessing exercises:

1. Plan and organize the evaluation
2. Observe the exercise and collect data
3. Analyze data
4. Develop a draft after action report
5. Conduct an exercise debrief meeting
6. Identify improvements to be implemented
7. Finalize the after action report
8. Track implementation

Exercise evaluation should be a formal process that involves input by all participants. For operations-based exercises, expert evaluators are used to observe the exercise play and note things that go poorly or well. Participants should also be asked to make note of any problems they encounter during exercise play. Immediately after the exercise, it is useful to conduct a quick debrief to capture ideas while they are still fresh in people's minds. This "hot wash" should be conducted by a skilled facilitator who seeks to identify what went well and what could be improved without fixing blame. Some jurisdictions use a questionnaire at the same time to capture ideas and numerical scores for the conduct of the exercise. Questionnaires sent home with participants, however, have a low rate of return.

Data from evaluators and from the after action debrief are analyzed and recommendations developed. Once this information has been compiled, it is useful to conduct a formal exercise debriefing with key officials to validate the findings and refine the recommendations. The expected outcome from the debriefing is a list of accepted recommendations and specific actions that will be taken. This information is then captured in an after action report and provided to the corrective action program for tracking.

Actual Incidents

Another form of assessment of capacity is the performance during actual events. Real events offer the best indicator of how a jurisdiction is likely to respond to crisis. Failure to implement plans or the failures of plans during implementation are all key indicators that there may be the need for additional training or planning. It is important to have a method for capturing what went well and what could be improved during actual operations.

NFPA 1600 provides an eight-step process that is useful in dealing with corrective action needs identified in actual events and correlates well with the HSEEP model for exercises:

1. Develop a problem statement that states the problem and identifies its impact.
2. Review the past history of corrective action issues from previous evaluations and identify possible solutions to the problem.

3. Select a corrective action strategy and prioritize the actions to be taken.
4. Provide authority and resources to the individual assigned to implementation so that the designated change can be accomplished.
5. Identify the resources required to implement the strategy.
6. Check on the progress of completing the corrective action.
7. Forward problems that need to be resolved by higher authorities to the level of authority that can resolve the problem.
8. Test the solution through exercising once the problem is solved.

The process for capturing and analyzing data is similar to that used in exercises: a quick debrief, analysis of reports and data, a formal debriefing, and the issuance of an after action report containing recommendations and specific actions to be taken. However, since these are real events, there may be other factors that limit the availability of data. For example, there may be legal concerns over the findings of the after action report and their impact on potential litigation. Departments may view the review of a response as an internal matter. While after-action reports generally do not apportion blame but merely seek to identify opportunities for improvement, there may be concerns that a negative finding may lead to disciplinary action. Consequently, the process for collecting and assessing information on real incidents needs to be a formal, documented process agreed to by all stakeholders.

Tracking Corrective Actions

During the National Emergency Management Baseline Capability Assessment Program (NEMB-CAP), the ongoing effort by DHS to assess states against the EMAP Standard, only a third of the states surveyed had corrective action programs. This highlights a major failing of emergency management programs across the country. It is not uncommon to find the same deficiencies reflected in after-action reports over and over with nothing every being done about them. Most emergency management programs do not have a documented system for tracking the implementation of corrective actions.

Once potential problems are identified through exercises, assessments against the EMAP Standard, or actual incidents, there should be a vetting process to determine whether the problem was the result of a unique set of circumstances or whether it requires a systemic change. If the problem is systemic, the task group that is reviewing the problem (e.g. the exercise evaluation committee) should develop recommendations that include items similar to those for performance objectives: statement of the problem, action agent, time line and milestones, associated costs, and source of funding.

The real issue is not the identification of problems or even determining corrective action. The real issue is tracking the implementation of the corrective action. This means developing a system that holds the action agents accountable for progress. Some recommendations may be appropriate for integration into the strategic plan and can be tracked through the annual review process. For others, it may be necessary to create a tickler file or database. However, these methods are ineffective unless they generate regular reports that are used as management tools by senior executives. Examples of these are a quarterly report provided to department heads, a monthly report provided to the advisory group, or an annual report provided to the senior elected official. Ultimately, the corrective action program must close the loop in a process that begins with assessment and ends with substantive improvements to the emergency management program.

CONCLUSION

An effective emergency management program is not a collection of various initiatives conducted under a variety of authorities and funded through disparate funding streams. It is a cohesive program that takes advantages of disparate elements to produce a synergistic reduction of risk and an increased capacity to respond. Program cohesiveness is developed and maintained through a strategic planning process that is inclusive and collaborative and fits the work of many agencies operating under separate authorities into a manageable framework. The program is documented through an administrative plan that provides governance and a multi-year strategic plan that lays out measurable performance objectives.

This chapter has also made the case for a change in the way emergency managers interact with local government. The old paradigm that held the emergency manager accountable for all issues related to crisis response is a recipe for failure. Instead, the proper role of the emergency manager is as an administrator with responsibility for coordinating the disparate elements of an enterprise-wide program that assists in the management of risk. In this context, the most important document produced by the emergency manager is not the emergency operations plan but the strategic plan that coordinates the actions of multiple agencies and groups toward a unified vision of community resilience.

Chapter 5

ASSESSING RISK

It is a capital mistake to theorize before one has data. Insensibly one begins to twist facts to suit theories, instead of theories to suit facts.
—Sherlock Holmes
From *A Scandal in Bohemia* by Sir Arthur Conan Doyle (1891)

Risk assessment is the foundation of the emergency management program. A properly done risk assessment allows program elements to be based on a realistic appraisal of the types of risks a community is likely to face. It allows the identification of scenarios based on realistic assumptions and provides justification for the commitment of program resources. It highlights opportunities for mitigation. In essence, it is the first step in the problem-solving model: a realistic definition of the problem.

In assessing risk, it is important to use a structured process to ensure that no potential threat to the community is overlooked. It is not uncommon for one to assume that the only serious hazards are the obvious risks that "everyone" knows about. It is also common to see crisis planning driven by what is known among emergency managers as the "flavor of the month": the particular threat that is currently the focus of the media and/or the public. The single biggest problem in risk assessment, however, is the common mistake of assuming that the hazard is the risk. A hazard is a dangerous event or circumstance that has the potential to cause an emergency or disaster. The hazard in itself may not necessarily present a risk to a community. It is the vulnerability of the organization to the hazard's impact and the probability of that hazard occurring that creates risk. Risk assessment is the analysis of these three factors: impact, probability, and vulnerability, at both the macro and micro levels.

HAZARD IDENTIFICATION

One begins the assessment of risk by first determining what hazards may affect the community and then ranking those hazards on the basis of how likely the hazard is to occur and the potential impact. Hazards tend to fall into two broad categories: naturally occurring hazards such as geological, meteorological, and biological hazards, and human-caused events that may be accidental or intentional. In general, natural hazards tend to be more widespread in their effects while human-caused events tend to have more localized impact.

This becomes apparent when one contrasts Hurricane Katrina with the attacks of September 11. One can certainly argue that September 11 had wide ranging impacts on national and world policy and the economy that may be difficult to quantify and may never be fully known. However, Katrina's current economic loss estimate of $125 billion dwarfs the General Accounting Office's estimate of $27.2 billion in losses from September 11. With the exception of loss of life (as of this writing, Katrina's death toll is just over 1,300 as compared to over 3,000 lost in the World Trade Center attack), Katrina exceeds the impact of the 2001 terrorist attacks in every sense: geographical area affected, number of evacuees, public resources committed to relief, displaced persons, cost of rebuilding, etc. The point is not to diminish the horror of the New York attack, but to point out that natural hazards pose a significant threat and that to focus on a single hazard, as has been the tendency in the U.S. since September 11, may actually increase one's vulnerability to other threats.

However, human-caused hazards cannot be ignored either. If one looks at potential threats beyond just terrorism, one can't help but note that many of these hazards are related to modern technological processes and are both more likely to occur and have a more rapid onset than natural hazards. Natural hazards may be seasonal or may have slow onsets, but commonplace events such as fires and hazardous materials spills can occur without warning. These events may lack the scale of a large natural event and may be very localized in scope, but for many communities such occurrences could well be catastrophic if they are unexpected and sufficient resources are not in place to deal with their consequences.

The community also needs to consider threats that may not necessarily constitute an emergency in the public sector but could have a severe impact on private businesses and the public. On December 8, 1998, an employee error caused a blackout to more than 375,000 businesses and households on the San Francisco Peninsula. For the emergency services, the event was largely an inconvenience: there was an increased need for traffic control at critical intersections and calls to free people trapped in elevators, but no real public safety issues. Power was restored within eight hours and the city of San Francisco recovered some $1.1 million in lost revenues and response costs. For the private sector, however, the outage resulted in major economic loss. The Pacific Gas and Electric Company paid out $8.3 million to settle some 6,600 private claims and was fined $440,000 by the California Independent System Operator, the organization that coordinates power resources in the state.

Power emergencies are examples of another type of hazard that should be considered: those that produce cascading effects. Cascading events are the catalysts that spark new disasters or extend the effect of the disaster well beyond the initially affected area. On August 14, 2003, parts of the eastern United States and Canada suffered the worst blackout in United States history. The blackout was caused when high voltage lines in Parma, Ohio came into contact with overgrown trees, triggering a shutdown of the system to prevent a short circuit. A combination of a computer bug that interfered with alarms and the high consumer demand in the summer season resulted in a sequential shut down of other systems throughout the region.

The blackout affected over 50,000,000 customers. Some areas lost water pressure because pumps were inoperative, resulting in potential contamination to the water systems and the issuance of boil-water

Figure 5.1 The 2003 blackout shut down public transportation and forced many New Yorkers to walk home. *Photo: Eric Skiff, www.glitchnyc.com.*

orders that lasted for up to four days. Sewer systems failed, releasing raw sewage into waterways and forcing the closing of ocean beaches. A chemical plant in Sarna, Canada accidentally released over 300 pounds of vinyl chloride into the St. Clair River. Transportation systems, including airports, were closed. Gas stations could not pump fuel, causing vehicles to be abandoned when they ran out of gas. Oil refineries were shut down, sparking a rise in gasoline prices. Cellular phone systems were inoperable. In some areas, the outages lasted until August 16 and full power was not restored to parts of Canada until August 19.

The 2003 North America blackout is an example of how a minor event (a high voltage line touching a tree) can produce a widespread problem that in turn generates other problems, such as water contamination, sewage releases, etc. This event points out the importance of the impact of an event rather than the event itself. In assessing the risk of a cascading event, the planner should not try to identify all the incidents that could lead to such an event but consider how a widespread event could affect the community. The lesson from the blackout is that a widespread power outage lasting several days is a potential hazard to the community. The duration of the outage and the impact on the community are variables that can be considered during an analysis of the hazard.

While it would seem to be a straightforward exercise, hazard identification can be complex. There is a tendency to focus on obvious or recurring hazards and to dismiss lesser hazards as unimportant. Often, decisions are made on the basis of incomplete or random information. There is also a tendency to be reactive and to prepare for the last disaster. The result is that communities are frequently surprised by events that were foreseeable but largely ignored because they did not have the visibility of the "popular" hazards.

An example of this was the sudden emergence of the Y2K hysteria in 1999. Computer programmers had known for years that computer systems recorded dates using the last two digits of the year and that the year 2000 had the potential for a massive failure of date-dependent systems. Warnings were sounded in the early 1990s, but few organizations made any attempt to take corrective actions, partly because of the cost and partly because few took the problem seriously. In 1999 the media picked up on the potential problem and generated a wave of public hysteria. The public was presented with a steady diet of doomsday scenarios that had aircraft falling from the skies, food distribution systems disrupted, nuclear power plants exploding, and society collapsing into anarchy. Even simple devices like microwaves would turn on the unsuspecting homeowner because of its imbedded microchip. Organizations and governments committed considerable funds to identify critical processes that might be affected by the loss of computer systems and to install fixes to potential problem. Yet the threat from the millennium bug was foreseeable and could have been mitigated well before 1999.

Another frequently overlooked threat is the impact of repetitive events that have a cumulative effect over time. An example of this is the series of rainstorms in California in 1998 resulting from the El Niño effect that caused widespread flooding. El Niño, or as it is more correctly called, the El Niño Southern Oscillation (ENSO), is a periodic disruption of the ocean-atmosphere system in the southern Pacific. A relaxing of the trade winds in the central and western Pacific reduces the efficiency of the ocean cooling system, resulting in an increase in the water temperature. One of the consequences of this warming trend is an increase in the rainfall in the southern tier of the United States.

In 1997, the National Weather Service predicted significant rainfall from the El Niño effect and the State of California conducted a series of briefings and workshops to help counties prepare for the storms. Resources were stockpiled and preventative measures put in place. Despite the best efforts of emergency managers throughout the state, the series of storms proved catastrophic. Individually, each of the storms would have had minimal impact on the state. However, the almost continuous bands of rain saturated the ground, resulting in increased runoff that produced flooding and mudslides throughout the state. The cumulative effect resulted in a presidential declaration of disaster involving 41 of the state's 58 counties. Nevertheless, the impact of the storms might have been much worse had the state not acted on the information from the National Weather Service.

Hazard identification should also include a consideration of the commercial processes that take place within a community. Sometimes these hazards are obvious, but in many cases they are so ingrained in the life of a community that it is easy to overlook them as threats. On January 15, 1919 a 50 foot tall tank of molasses exploded in the North End Park area of Boston, probably as the result of overfilling and fermentation brought on by an abrupt and unseasonable increase in the temperature. The estimated 2.5 million gallons of molasses created a wave between eight and 15 feet in height moving at 35 miles an hour, exerting a force of over two tons per square foot. The wave killed 21 people and injured 150 and the cleanup took over six months.

In identifying hazards, then, it is important to consider all the threats that could have an impact on the community. Many of these can later be eliminated or given low priority in the analysis phase but it is critical to make sure that nothing has been missed. On the other hand, a

Figure 5.2 Cumulative rainstorms from El Niño produced widespread flooding in California in 1997. *Photo: Robert A. Eplett/OES CA, California Governor's Office of Emergency Services.*

healthy dose of common sense is required. One can spend endless hours concocting scenarios and playing the "what if" game without ever completing the hazard identification. One should keep in mind that the ultimate goal of hazard identification is to consider a sufficiently broad range of threats to ensure that all potential impacts have been identified. The emphasis is not so much on the type of hazard as it is on the impact of the hazard. For example, an electrical outage can be caused by any number of hazards. What is important in hazard analysis is to determine as much as possible the potential maximum duration and extent of an electrical outage.

Hazard identification has never been easier. Many of the potential threats affecting specific areas of the United States have been identified and the Internet provides almost unlimited resources for research. Government agencies such as the National Weather Service, the U.S.

Figure 5.3 The Boston molasses disaster leveled several blocks of Boston in 1919. *Photo: Leslie Jones, Boston Herald, 1919.*

Geologic Survey, and the Federal Emergency Management Agency are good starting points for collecting data on potential hazards. FEMA manages programs such as the National Earthquake Hazard Reduction Program (NEHRP), the National Flood Insurance Program (NFIP), and the National Dam Safety Program (NDSP) that make data available, either via the Internet or at a nominal cost. There are also a number of computer models available at little or no cost, such as FEMA's Hazards United States (HAZUS) loss estimation tool that offers models for hurricanes, floods, and earthquakes. FEMA is also coordinating a multi-hazard mapping initiative that is intended to make hazard maps from various sources available on a single web site.

Over the years, various pieces of legislation have broadened the rights of a community to know the potential human-caused threats that might affect it. For example, SARA Title III and the Clean Air Act established reporting requirements for an extensive list of hazardous materials that can be accessed by the general public. OSHA has requirements related to the posting of Material Safety Data Sheets in the workplace that identify the types of chemicals used, their potentially harmful effects, and the treatment for exposure. The Environmental Protection Agency (EPA) administers a Toxic Release Inventory that tracks the release of toxic chemicals from manufacturing companies by zip code and offers access to databases related to environmental hazards.

In gathering data related to hazards, one must be careful to understand the nature of the data being examined. It is a common practice in hazard analysis to use information products developed for a special purpose to assist in providing general information about the nature of a threat. However, this data may have been assembled for a very specific purpose that may make the information of limited value. The data from such sources may be useful for approximating probabilities and determining scope but it is essential that the original purpose of the data and its limitations be understood. However, many planners make assumptions about information products and use them for purposes for which they were never intended.

An example of this is the use of Flood Insurance Rate Maps (FIRMs) to develop emergency plans. To understand why this is a problem, one needs to understand the original purpose behind the development of FIRMs.

The most common disaster in the United States is flooding. To assist in mitigating this threat, the federal government created the National Flood Insurance Program (NFIP) to encourage the adoption of flood hazard reduction ordinances in exchange for subsidized flood insurance. In exchange for adopting a model flood ordinance, citizens of a participating community are allowed to purchase flood insurance subsidized by the government. As negative incentive, federally insured mortgages required flood insurance in flood hazard zones, essentially slowing growth in non-participating communities. The NFIP currently serves over 4.5 million policyholders and provides $650 billion in coverage.

The rates for the insurance are based on FIRMs. FIRMs are used by insurance agents to determine policy premiums and by banks to determine the requirement for flood insurance when considering a federally insured loan. The agent or banker determines the location of the structure in question on the FIRM and determines the flood plain category, if any, in which it falls.

FEMA is responsible for compiling, updating, and distributing FIRMs. This involves maintaining nearly 100,000 separate map sheets for over 21,000 communities across the United States. The volume of maps is enormous: FEMA keeps over 40 million maps in stock and prints some six million separate panels each year. The large scale of the maps (1:4,800 to 1:24,000) and the limited information on them (they contain only transportation networks and waterways, not topographical features) make them uninformative for any other use. They are too numerous to be included in the Depository Library Program and FEMA has not made them available online as of this writing. Consequently, maps can only be ordered from FEMA or viewed at a local planning office. (FEMA is in the process of a major modernization program to convert the program to digital maps that can be updated quickly and be easily accessible. The expected project completion date is in 2010.)

Estimating the base flood level that forms the core of the maps is a complex undertaking. Base floods are determined by mathematical computations involving frequency and risk. In theory, this involves the analysis of discharge and velocity over a 100 year period, data that may be unavailable or questionable. In addition, special analyses are required to factor in variables such as ice jams, alluvial fans, dams and levees, and hurricanes. The complexity of the process means that maps are frequently out of date and inaccurate. This inaccuracy results in local communities frequently challenging the maps. Changes generated in this way can take up to a year to be reflected in a change to the appropriate FIRM.

Nevertheless, there is a tendency among some planners to treat FIRMs as if the data on them is complete and accurate and to use them for crisis planning, such as determining evacuation routes. This demonstrates a lack of understanding that a FIRM is merely a representation of a complex dynamic process, designed for a single, specific use: setting insurance rates. The FIRM was never intended to be a topographical tool to assist in operational planning. It lacks the important details that would be needed for such a purpose and is an estimate rather than a precise measure of where floodwaters may go. As an actuarial tool, FIRMs are essential for the implementation of the National Flood Insurance Program. As a hazard assessment tool, they are merely an indicator that the community is prone to flooding and should not be used for emergency planning.

HAZARD ANALYSIS

Once potential hazards that could affect an organization have been identified, the logical question follows, "Which ones should I be concerned about?" The answer can be found by analyzing the probability and impact of each disaster. In a sense, though, the question is not precisely the right one. The real issue, as noted above, is not the type of disaster, but the consequences. In assessing risk, one is trying to determine the impacts of a range of hazards to arrive at a reasonable basis for developing a risk management strategy.

"Impact" is a relative term based on the vulnerability of the community to a specific hazard. In assessing impact, one considers the agent-specific effects of the hazard and the community's resilience in the face of those effects. One also factors in the risk tolerance of the community. For example, where communities confront the same hazard on a regular basis, there is a certain amount of adaptation that may take place. An event that is viewed as a disaster to other communities may be considered an inconvenience in such an adapted community.

In the simplest form of hazard analysis, threats can be analyzed by ranking them on the basis of a low, medium, or high probability or impact. The results can be shown in a matrix such as that in Figure 5.4, sometimes referred to as a vulnerability analysis worksheet. By itself the worksheet doesn't provide a lot of detail and represents a minimum effort at analysis. Displaying the information in graphic form (Figure 5.5) provides a better indication of the relationships among the hazards.

A slightly more sophisticated approach is to assign a numerical value to each factor. For example, high = 3, medium = 2, and low = 1. This allows the graph to be more precisely plotted. Figure 5.6 is an example of what Steven Fink, author of *Crisis Management: Planning*

Hazard	Probability of Occurrence	Potential Impact	Remarks
Earthquake	High	High	70% probability within 30 years
Tsunami	Low	High	City Hall located within run up zone
Terrorist Attack	Low	Medium-High	Three potential targets on State list
HAZMAT Spill	Low	Medium	
Flood	Low	Low	No critical infrastructure in flood zones

Figure 5.4 Hazard analysis matrix.

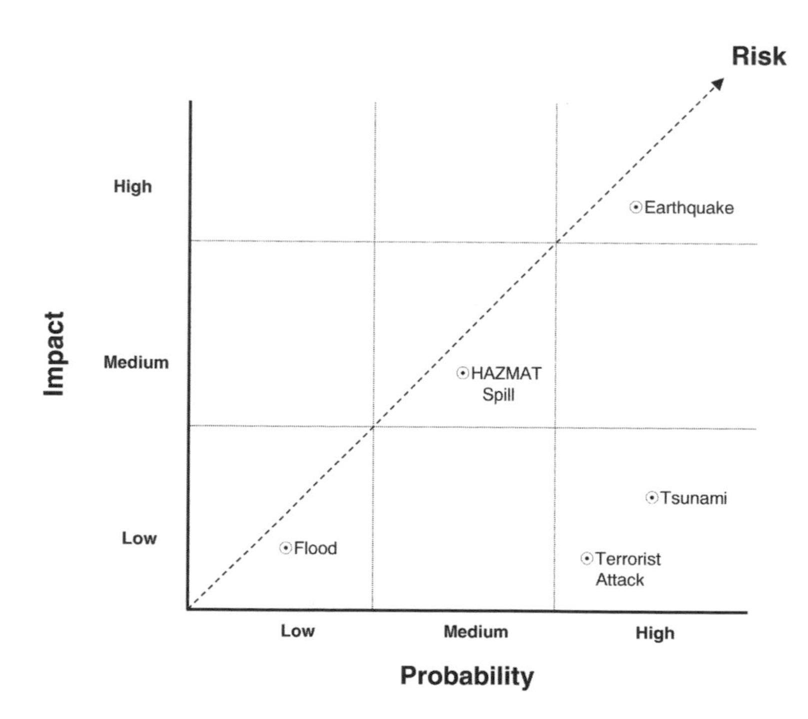

Figure 5.5 Hazard analysis graph.

for the Inevitable, refers to as the crisis barometer. The crisis barometer determines the relationship between probability and potential impact for a hazard using a scale of 1 to 100 percent for probability and a scale of one to ten for severity of impact. This relationship is then used as a polar coordinate on a graph to assign the hazard to one of four zones of priority. The result is a visual representation of the various risks facing a community and their relationship to each other.

This simple ranking provides a quick way of qualitatively grouping the hazards and can be useful in reducing the range of hazards that must be analyzed. For example, the decision may be made to drop those hazards rated low probability and low impact and to focus on the remainder with priority on those that have high impact and high probability. However, it provides little detail on the specific impact of the

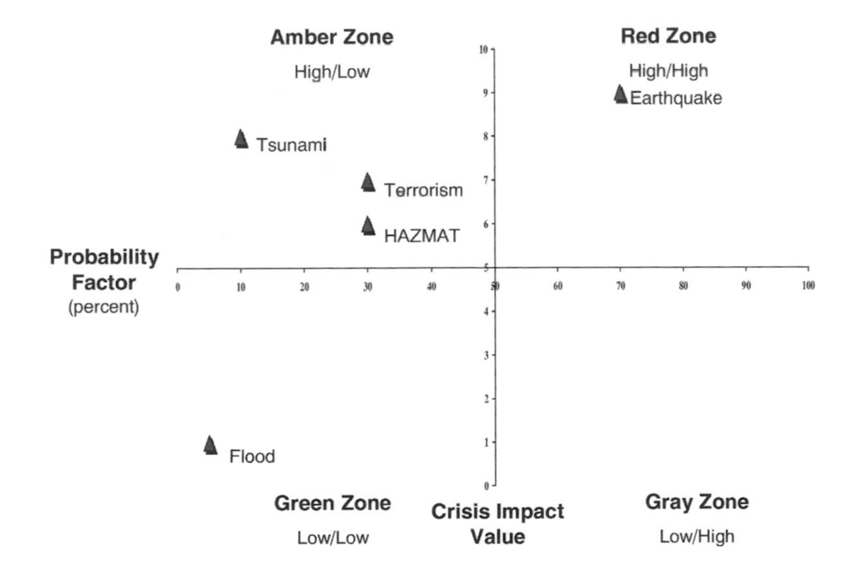

Figure 5.6 Crisis barometer. Based on Steven Fink, *Crisis Management: Planning for the Inevitable*.

hazard on the community and should be considered only a starting point in the analysis process.

The ranking of a hazard on the basis of its impact is, to a certain extent, subjective in that it is based on the community's tolerance for risk. What may be an acceptable level of risk for one community may be catastrophic for another. It is important, therefore, to agree up front on the specific definitions of terms to be used in the analysis (e.g., what do we mean by low risk?). This process allows for the inclusion of this subjective risk tolerance by defining acceptable levels of risk in terms of the specific community. Once these definitions are in place, hazards can be assessed with more objectivity using the agreed upon criteria to produce more consistent results.

A very simple approach to this type of analysis is shown in Figure 5.7. Each hazard is ranked on a scorecard as to its expected impacts. The list in Figure 5.7 is representative only and could be replaced by any number of factors the organization considers important. In this simplified methodology, if the hazard produces an impact on the target factor, it is checked off. The total number of checks is then scored using the scale to produce a result of low, medium, and high. There should also be some consideration given to a definition of these terms such as low = handled with local resources, medium = requires state assistance, and high = requires federal assistance.

Figure 5.8 is an example of the hazard profile worksheet used by the Federal Emergency Management Agency in its independent study course IS235 Emergency Planning. In this example, FEMA uses four criteria for measuring impact and four criteria for probability. In each case, the individual criteria are defined, providing guidance for ranking the hazards in those categories. One prepares a profile for each threat and then uses the profiles to compare and rank order the hazards. The information on the profile sheet can also be transferred to a vulnerability analysis worksheet to make this comparison easier.

The use of defined criteria can be used to further expand the analysis of the hazard to consider areas of specific concern to the community. Figure 5.9 is an example used by FEMA to rank the potential impacts of a hazard in three areas critical to public sector response: casualties, loss of critical facilities, and property damage. Again, the pre-identified criteria allow for a relatively simple ranking of the impacts of the hazard and help to ensure consistency when analyzing

Hazard	
Potential Impact	
Injuries	
Fatalities	
Public Health Issues	
Property Damage	
Economic Loss	
Essential Services Disrupted	
Critical Infrastructure Shut Down	
Score	

Rating		
Level	**Score**	
Low	0-3	
Moderate	4-6	
High	7-10	

Figure 5.7 Hazard profile.

multiple threats. Figure 5.10 combines the criteria from both FEMA products to produce a single matrix. This matrix could be used to rank each hazard in each of the areas of impact identified on the matrix.

The FEMA criteria used above are only representative. Ideally, the community will determine what vulnerabilities are important within the context of its risk tolerance. NFPA 1600 requires that hazards be analyzed with respect to their impact on nine key areas:

Hazard Profile Worksheet

Hazard:

Potential magnitude (Percentage of the community that can be affected):

- ☐ Catastrophic: More than 50%
- ☐ Critical: 25 to 50%
- ☐ Limited: 10 to 25%
- ☐ Negligible: Less than 10%

Frequency of Occurrence:	Seasonal Pattern
•Highly likely: Near 100% probability in next year. •Likely: Between 10 and 100% probability in next year, or at least one chance in next 10 years. •Possible: Between 1 and 10% probability in next year, or at least one chance in next 100 years. •Unlikely: Less than 1% probability in next 100 years.	

Areas Likely to be Affected Most

Probable Duration:

Potential Speed of Onset (Probable amount of warning time):

- Minimal (or no) warning.
- 6 to 12 hours warning.
- 12 to 24 hours warning.
- More than 24 hours warning

Existing Warning Systems:

Does a Vulnerability Analysis Exist?
Yes
No

Note that some hazards may pose such a limited threat to the community that additional analysis is not necessary.

Figure 5.8 Hazard profile worksheet. FEMA IS235 Emergency Planning.

1. Health and safety of persons in the affected area
2. Health and safety of responding personnel
3. Continuity of operations
4. Property, facilities, and infrastructure
5. Delivery of services
6. The environment

Potential Impacts			
Level	***Casualties***	***Critical Facilities Shut Down***	***Severely Damaged Property***
Catastrophic	Multiple deaths	+1 month	+50%
Critical	Permanent disabilities	+ 2 weeks	25-50%
Limited	Temporary injuries	1-2 weeks	10-25%
Negligible	Treatable w/1st aid	-24 hours	-10%

Figure 5.9 Defining levels of impact. FEMA IS235 Emergency Planning.

Level	Hazard Information		Human Impact		Economic Impact		Service Impact	
	Frequency of Occurrence	Speed of Onset	Population Affected	Casualties	Property Damage	Economic Loss	Critical Facilities Shut Down	Essential Services Disrupted
Catastrophic	100% within 1 year	Minimal	>50%	Multiple deaths	> 50%	>$ 1 billion	>1month	>1month
Critical	10-100% within 10 years	6 to 12 hours	25-50%	Permanent disabilities	25-50%	$100 million - $1 billion	> 2 weeks	> 2 weeks
Limited	1-10% within 100 Years	12-24 hours	10-25%	Treatable injuries	10-25%	$10 – 100 million	1-2 weeks	1-2 weeks
Negligible	<1% within 100 years	>24 hours	<10%	Treatable w/1st aid	<10%	<$10 million	<24 hours	<24 hours

Figure 5.10 Impact analysis matrix.

7. Economic and financial conditions
8. Regulatory and contractual obligations
9. Reputation or confidence in the organization

These nine factors represent a minimum standard and other factors important to the community can be added. The criteria for determining whether an impact is high, medium, or low should be defined for each factor based on the risk tolerance of the community.

One area required by NFPA 1600 that is often not considered by most communities during the hazard analysis is the impact of a potential crisis on the reputation of the affected organization or community. This is clearly a subjective determination, but a crisis that affects the reputation of an organization can have far ranging consequences, for good or ill. The sidebar case study illustrates an example from the private sector as to how appropriate response can enhance reputation. It also highlights the damaging effect of a poorly thought out response to crisis.

The public sector is not immune to loss of reputation. Following Hurricane Hugo in 1989, Senator Ernest Hollings (D-SC), in criticizing FEMA's slow response to the hurricane, referred to the agency as "…the sorriest bunch of bureaucratic jackasses I have ever encountered in my life." The image of Kate Dale, emergency director for Miami-Dade County, demanding, "Where's the cavalry?" during Hurricane Andrew in 1992 garnered national attention and tarnished the agency's reputation even further. Hurricane Katrina in 2005 brought similar claims of federal incompetence and diminished the reputations of the mayor of New Orleans and the governor of Louisiana. In contrast, the State of Florida was widely praised for its aggressive handling of hurricane response in 2004 and 2005 and Mayor Rudi Giuliani has built a successful career as a speaker and consultant because of his leadership during the September 11 attack on New York.

Once profiles are prepared for each hazard, they can be compared by combining the results in a matrix or worksheet. Figure 5.11 shows a partial worksheet based on one developed by a major university for use in its business continuity program. Note that this example uses a numerical scale rather than defined criteria. The use of a scale is not as precise as using defined criteria as it requires a subjective decision on

Type of Emergency	Probability	Human Impact	Property Impact	Business Impact	Internal Resources	External Resources	Total
Fire	2	3	5	4	4	1	19
Flooding	2	1	4	3	4	3	17
Severe Storm	3	3	3	3	4	1	17
Terrorist Attack	1	4	4	1	4	1	15
Short Term Power Outage	1	1	1	1	2	1	7
Long Term Power Outage	1	1	1	1	2	1	7
	High Low 5-----1	High Impact 5-----		Low Impact -----1	Weak Resources 5-----	Strong Resources -----1	

Figure 5.11 Impact analysis worksheet.

the part of the planner. It has also been suggested that the use of a scale, particularly one having an odd number of rankings, may result in the answers being skewed to the mid-range of the choices.

The example in Figure 5.11 also introduces the use of numerical weighting in assessing hazards. In this case, the use of the numerical scale provides a cumulative figure that can be used to rank the hazards. If one combines a numerical weighting with defined criteria, it is possible to further refine the analysis to conform with the community's tolerance for risk by providing extra weight to those factors the community considers more important. For example, loss of life might be given a higher factor than severe facility damage.

The danger with using a numerical system, however, is that it can be perceived as somehow more "precise" because of its use of numbers. This creates a tendency to focus on the highest ranking threat rather than on a range of threats. For most purposes, the use of numerical weighting does not really offer much advantage over the use of defined criteria (which should consider the same subjective factors that the numerical system seeks to quantify). This is because hazard

CASE STUDY: TYLENOL AND PERRIER—REPUTATIONS AT RISK

In 1982, the Johnson and Johnson Company was faced with a series of deaths traced to cyanide-laced capsules of its pain reliever Tylenol. The company responded with a media campaign that made it clear that customer safety was the company's primary concern. It immediately notified consumers of the threat, stopped advertising and development of the product, and recalled 31 million bottles of Tylenol capsules at a cost of more than 100 million dollars. The company even went so far as to offer consumers who had already purchased Tylenol capsules the opportunity to exchange them for tablets at no cost. These actions were in stark contrast to the normal approach of other corporations of cost containment and self-protection and reflected Johnson & Johnson's corporate values that put the customer first. Johnson & Johnson won widespread praise for their reaction and, through the use of a well-crafted media campaign, rebuilt Tylenol's reputation with consumers.

The proactive approach of the Johnson & Johnson Company stands in stark contrast to how other companies have handled issues of product contamination. In 1990, regulators in North Carolina reported that they had found the carcinogen benzene in Perrier bottled water. Source Perrier initially offered various explanations for the contamination, finally stating that it had traced it to an isolated employee error that affected bottles used for North America. The company recalled a small amount of product in the United States and Canada, amounting to some 70 million bottles of water (Perrier sales at the time were 1.2 billion bottles worldwide). A short time later, officials in Holland and Denmark reported finding similar benzene contamination in Perrier product in Europe, sparking a worldwide recall of the product. Source Perrier changed its explanation yet again. The company now claimed that benzene was a normal byproduct of carbon dioxide that was usually filtered out during the bottling process. It again blamed employees, this time claiming that they had not changed filters. Closely following this crisis was the disclosure the same year that Perrier artificially carbonated its water and the U.S. Food and Drug Administration forced it to remove the words "Naturally Sparkling" from its product. Prior to the benzene crisis, Source Perrier was the leading importer of bottled water in the U.S., holding some 6 percent of the market share. By 1995, Perrier sales had dropped by 50 percent and Perrier is no longer a key player in the U.S. bottled water market.

analysis is ultimately a subjective process. Hazard information based on estimates and projections is measured against vulnerability in the context of the community's tolerance of risk. It is never more than an estimate, no matter how carefully or precisely done. This is why emergency planners should be more concerned about potential impacts than specific hazards.

To summarize, hazard analysis ranks the potential threats facing a community by considering the relationship between probability and impact for each hazard. Threats found to have low probability and low impact can be eliminated from further analysis while a more detailed analysis is performed on the remaining hazards. This analysis begins with a definition of the criteria that will be used to assess the hazard based on the risk tolerance of the organization. The analysis is usually displayed in summary form, such as the hazards vulnerability worksheet.

The end result of the hazard analysis should be a realistic assessment of the potential impacts of various hazards on the community. Again, note the emphasis on the impact. It is impossible to say for certain what hazards will occur but one can make a fairly accurate assessment of the impacts that an emergency management program will need to confront. Is it likely that key facilities will be destroyed? If not, are they likely to be isolated? What critical services are likely to be disrupted and for how long? Will key employees be able to get to work?

One final thought on the process of hazard analysis: one cannot discount the value of intuition. The process just described is reasonable and based on proven methodologies. Estimates of probability, which form the basis of this system, are the result of statistical estimations based on observed data. However, catastrophic events have probabilities that are far below those which are statistically observable. Consequently, the type of modeling that may work for disasters does not hold for larger scale events and one is forced to conclude that, when dealing with catastrophic events of infrequent occurrence, probabilities must be discarded in favor of impacts. In other words, although the incidence of occurrence is extremely low, the impact is so devastating that a jurisdiction must plan for it. This suggests that there must be a subjective determination made as to how credible a high-impact event might be by considering the possible rather than the probable.

Credibility is subjective but there are tools on which the determination can be made. One can consider past events, either through the historical record, or through scientific observations such as geological or meteorological records. One can also look at scientific projections or estimates. However, rather than the precise models that one is used to in disasters, one may be forced to rely on an ensemble of models that represent possibilities rather than projections. Finally, one can use a good dose of common sense. The catastrophe in New Orleans in 2005 was predictable and had been the subject of numerous articles and reports. A similar situation exists in the delta of the Sacramento River in California where an aging levee system threatens not only local communities but the water supply for southern California as well.

IMPACT ANALYSIS

Hazard identification and analysis provide a macro view of potential disasters. That is, they provide qualitative assessments of the types of impacts the organization or community may face. They suggest an order of magnitude or scope and identify the services or infrastructure that might be disrupted or destroyed. However, this information does not address the impact of the hazards on individual agencies except in the broadest sense. For this micro view, one must turn to the impact analysis or business impact analysis, as it is known in the private sector. Where hazard analysis looks at impacts on the community as a whole, impact analysis considers the impact of a hazard on specific departments or agencies.

Impact, or business impact analysis (BIA), is the identification of the effect of risks on the organization as a business. BIA identifies those critical functions and processes that must be continued during crisis if the organization is to be able to function. Identifying these critical functions and giving them priority for planning and resource allocation allows for more focused and cost effective crisis planning. As such, it is pertinent to both the public and private sectors.

There is an unspoken assumption in most public sector response plans that departments and agencies will continue to function in support of the plan. This may be the case in the short term where the issue is immediate life safety operations, but governments are ultimately businesses as well. They must accumulate and disburse revenue, meet contractual obligations, pay employees, and deliver services. Disasters disrupt this process for government in the same way it does for business. For example, a reduced inflow of tax revenue means curtailed services at precisely the same time the community needs them the most. Following Hurricane Katrina, the City of New Orleans was forced to lay off 3,000 government workers, about half the work force. The same problem occurred following the San Francisco earthquake in 1906, creating problems in dealing with criminal activity during the reconstruction because of a reduced police force. Consequently, continuity of operations (COOP) and continuity of government (COG) planning are critical in enhancing a community's capacity to respond to crisis.

The concept of ensuring continuous government operations dates back to at least 1982 with the issuance of National Security Decision Directive 55 Enduring National Leadership dated September 14, 1982. The most recent iteration, Presidential Decision Directive 67, Enduring Constitutional Government and Continuity of Government Operations, dated October 21, 1998, establishes Continuity of Operations (COOP) as a unifying concept to ensure the continuance of essential government functions.

Federal Preparedness Circular 65 Federal Executive Branch Continuity of Operations (COOP) was reissued on June 15, 2004 to provide more defined criteria for COOP. As part of these criteria, FPC 65 establishes the requirement for access to a local area network (LAN) and vital files, records, and databases within 12 hours of COOP plan activation. In addition, the COOP plan must provide for sustained operations for up to 30 days until normal business activities can be reconstituted.

While there are no specific requirements mandating that local governments do COOP planning, it is an implied task. FEMA has stated on its web site that "...the development and maintenance of a viable COOP plan and capability at each level of government/jurisdictional responsibility is critical to save lives and protect the public health and well-being, protect property and preserve assets, maintain functionality, and maintain basic government operations and services."

Further, the EMAP standard requires that each agency identified in a jurisdiction's emergency plan have a COOP plan. Nevertheless, COOP planning is still very much in its infancy and the NEMB-CAP data showed only a 14 percent conformance rate with the EMAP Standard.

The starting point for COOP and COG planning is the BIA. The closest that the public sector has come to performing a BIA was the preparation for the Y2K crisis in 1999. For the first (and, unfortunately, probably the last) time, public agencies began to ask questions such as:

- What are my critical functions and processes?
- What critical processes are likely to be disrupted and for how long?
- How will disruptions in the supply chain affect me?
- What outside services do my critical processes depend on?
- In the event of wide scale disruption, what processes should I restore first?

Where the impact analysis could be performed by a small number of planners, the BIA is an enterprise-wide project and should be organized as such. The project team will require support from, and access to, senior executives and officials as well as supporting staff. The BIA project is normally organized into five phases:

1. Project Planning
2. Data Collection
3. Data Analysis
4. Report Presentation
5. Reanalysis

Project planning establishes the scope and methodologies for the project. Scope is particularly critical in conducting a BIA. As in hazard analysis, there can be a tendency to over analyze and to spend considerable time and resources without reaching any conclusions. In an initial BIA, one should focus on management level functions and critical interdependencies without trying to cover all sub-processes. For example, one can state that a particular process must be online within three days without describing all the sub-processes that must be performed to reach this objective. These sub-processes can be addressed in later planning or subsequent analyses.

Another critical part of project planning is the selection of a methodology for gathering data. While it may seem possible to intuit critical functions and their co-dependencies, doing so assumes a complete and total knowledge of all processes. Just as in approaching hazard identification, there is a risk that a critical process may be ignored or overlooked. Therefore, the BIA must follow a structured process that allows for results based on facts and not just informed guesses.

In *Business Continuity: Best Practices*, Andrew Hiles identifies nine common methods for collecting data:

1. Critical Success Factor Analysis—this method involves identifying areas of the organization's operations that must continue during crisis or have a significant impact on the success of the organization. For example, the ability to collect revenue or to provide specific services may be critical success factors. Once the critical success factors are identified, one then identifies the business processes that support them.
2. Key Performance Indicator Review—key performance indicators are statistics used to measure success in achieving goals and objectives. Processes that produce these measures are critical to the organization successfully meeting these goals and objectives.
3. Process Flows—charting process flows is a useful way of determining dependencies on other processes, such as supply chains. This method is particularly useful when coupled with another type of analysis known as Failure Mode and Effects Analysis (FMEA). FMEA looks at a system and considers the effects on the system if one or more elements were to fail.
4. Outputs and Deliverables—this method works backwards from expected outcomes to trace the processes needed to deliver those outcomes. Like the process flow method, it is useful for determining dependencies. A similar method is Fault-tree Analysis, in which an undesirable event is identified rather than an expected outcome and the potential causes of this undesirable event are identified.

5. Activity Categorization—in this method, processes are divided into three categories: core processes, support processes, and discretionary processes. This allows the broad prioritization of processes and allows for planning to be focused on processes important to the organization.
6. Document Review—this method relies on the review of key documents such as strategic plans, procedures and standards, insurance policies, contracts, and technical documents to determine what the organization considers important. This method seeks to identify factors that will help focus and prioritize planning such as key products, services, and customers and to quantify potential losses.
7. Questionnaires—a frequent method for collecting data is to develop a questionnaire that is distributed to key managers. Providing there is strong support from senior officials, this method can be useful in assembling an initial estimate of critical processes. However, it usually is not sufficient unless used in conjunction with other methods or as a starting point for interviews.
8. Interviews—interviews with key personnel and managers are important as both an initial instrument to collect data and as a reality check for assumptions and conclusions. However, they must be structured rather than free flowing. Interviews generally use a standardized list of questions that may mirror or expand on a previously distributed questionnaire. The use of this standard approach ensures that all interviewees are asked the same questions and reduces the risk that an important issue will be overlooked.
9. Workshops—workshops capitalize on the synergy of the group environment. Workshops are useful for gathering input from multiple respondents that might not otherwise be interviewed due to time or budget constraints. They can also be used to identify candidates for more in-depth interviews. Like all meetings, the workshop must be carefully facilitated to create a productive environment where participants are free to speak their minds and the meeting focus is maintained.

Using several of these methods of data collection, the planner can identify systems and processes that are essential to the organization's business. The planner can also make some estimate of what the impact of the loss of these systems and processes might be. The impacts may be either quantitative or qualitative. Quantitative losses are those that can be measured or extrapolated based on the data. For example, one can estimate the loss of revenue generated from parking fines because traffic control officers are reassigned to emergency traffic control. Note that for the purpose of the BIA, these figures do not have to be precise calculations—whether the loss is estimated at $1 million or precisely calculated at $999,527 per day is immaterial.

Qualitative loss estimation is more subjective. It may not be possible to calculate a precise figure for a projected loss or such calculations may be too speculative to be useful. Yet there are still losses that will have a major impact on the organization and need to be considered in any assessment of risk. One example that has been previously discussed is the effect of crisis on reputation. While it may be difficult to quantify the loss of reputation, one can certainly describe such a

Tangible Impact					
Business Function	**Department**	**Priority**	**RTO** (in hours)	**$ Loss per day**	**Comments**
Order entry	Customer Service		4	$500,000	20% of product accounts for 80% of revenue
Billing	Customer Service		8	$25,000	
New Account Credit Checking	Customer Service		16	$20,000	
New Account Set Up	Customer Service		16	$20,000	
Order Picking and Shipping	Distribution		4	$500,000	
Product Receipt and Storage	Distribution		72	$100,000	Sufficient stock of all products for 3 days
Purchase Order Creation	Finance		120	$10,000	
Accounts Receivable	Purchasing		72	$100,000	

Figure 5.12 Business impact analysis worksheet—quantitative assessment, Canadian Centre for Emergency Preparedness.

Intangible Impact					
Business Function	Department	Priority	RTO (in hours)	Impact High (3) Med (2) Low (1)	Comments
Product Recall	Customer Service		8	3	Would look incompetent – open to lawsuits
Customer Follow Up Surveys	Customer Service		240	1	
Payroll	Finance		4	3	Disgruntled employees – key staff might leave
Accounts Payable	Finance		120	1	
General Ledger Creation/Maintenance	Finance		120	1	
Employee Performance Reviews	Human Resources		240	1	
Marketing Literature	Marketing & Sales		72	2	

Figure 5.13 Business impact analysis worksheet—qualitative assessment, Canadian Centre for Emergency Preparedness.

loss qualitatively. Loss of public confidence in its administration can have long-range effects on the ability of a community to recover from a crisis.

In conducting the BIA, one must consider not only the initial impact of a particular risk but also its effects over time. There are two principal reasons for this. First, the longer an event continues, the more stress is placed on the organization as resources are expended and personnel become exhausted. The second reason is really a corollary to the first: not all critical processes may need to be reinstated immediately. Consequently, the BIA seeks to determine not only critical processes but also a factor known as the Recovery Time Objective (RTO). The RTO is the time between the loss of the process and when that loss begins to affect the organization (i.e. the longest time that the organization can afford to be without the system or process). This is sometimes referred to as the Maximum Acceptable Outage (MAO). Determining this factor becomes crucial as one begins to construct a continuity plan as it will assist in determining which processes must be addressed first.

Analyzing the data accumulated during the BIA should lead to several key results. The analysis should yield:

- An identification of systems and processes critical to the organization's survival as an organization.
- An estimate of the impact of losing critical systems and processes in qualitative and quantitative terms.
- An understanding of the interdependencies among and between the different core processes and the processes that support them.
- An estimate of Recovery Time Objectives for critical systems and processes.

The results of a BIA can be displayed in matrix form in much the same way that the results of the impact analysis were. Figures 5.12 and 5.13 show two examples of a business impact analysis worksheet. The first considers the tangible impact on business functions (quantitative impacts) while the second rates intangibles such as loss of reputation (qualitative impacts).

CONCLUSION

Hazard identification, hazard analysis, and business impact analysis are individual components of an overall assessment of risk for the organization. However, they do not represent stand-alone processes but are instead a continuum that progressively narrows a rather wide universe of potential hazards to focus on risks that are specific to the organization or community. They represent a structured process aimed at gaining a realistic appraisal of those risks and an understanding of how those risks will affect the organization. The risk assessment is the problem definition that will ultimately drive the development of the emergency management program strategy and the development of crisis-related plans.

Chapter 6

DEVELOPING STRATEGY

Strategy without tactics is the slowest route to victory. Tactics without strategy is the noise before defeat.
—Sun Tzu

As was noted in the previous chapter, risk assessment is the basis of the emergency management program. Upon completion of a risk assessment, planners should have a fairly complete picture of community vulnerability to potential hazards. Risk assessment also helps to put hazards in perspective, allowing for policy decisions that prioritize the use of scarce resources and focus planning on those events that represent the greatest risk to the community.

Basing emergency management policy on risk would seem to be common sense but the record of government has not been particularly good in this area. Funding for emergency management through the Emergency Management Preparedness Grant has always used a fixed base amount for each state plus a small amount based on population. When funds are looked at on a per capita basis, there is a great inequity in the distribution of funds, with those states with the greatest risk actually receiving less per capita than smaller states. Further, there is no basis for believing that population alone is a reasonable indicator of potential risk.

When one considers the much-trumpeted war on terrorism, one is forced to conclude that this policy is also not based on an accurate assessment of risk. If one looks at terrorism from an actuarial perspective, the mortality rate from terrorism in 2001 was on a par with deaths from drowning (2,978 to 3,247). In contrast, 700,142 Americans died of heart disease that year and the common flu kills about 36,000 a year. The economic impact of September 11 is estimated at $84–92 billion; the cost for Hurricane Katrina is estimated to exceed $200 billion. This is not meant to in anyway diminish the events of September 11 or to suggest that terrorism is not a threat. It merely points out that a community's risk to terrorism is a product of its vulnerabilities and the risk will vary from community to community, as it does with any hazard. Terrorism is not the principal risk in every jurisdiction in the United States.

Basing emergency management policy on risk may actually require an adjustment for many jurisdictions. Most emergency management programs are the product of the Cold War, with their single focus on national preparedness and additional responsibilities have accrued to them over the years without a basis in local policy. However, if the emergency management program is to be cohesive and consistent with community values and goals, it must be driven by strategies derived from policies based on risk.

A NEW LOOK AT AN OLD MODEL

One of the seminal events in the history of emergency management was the National Governors' Association study in 1978 that introduced the comprehensive emergency management model. The model recognizes four phases of emergency management:

- Mitigation—efforts taken to eliminate or reduce the impacts of hazards
- Preparedness—efforts to develop the capacity to respond to disasters
- Response—actions taken to deal with the impact of a disaster
- Recovery—actions taken to restore the community to normal

Figure 6.1 Comprehensive emergency management model.

Since the advent of Homeland Security, there has been a push by law enforcement to add a fifth phase called prevention. The draft changes to NFPA 1600 adds prevention and defines it as "activities to avoid an incident or to stop an emergency from occurring." It further limits mitigation to "activities taken to reduce the severity or consequences of an emergency." This type of political hairsplitting serves little purpose as the model has successfully formed the basis of emergency planning for almost 30 years. The definition of mitigation as "efforts taken to eliminate or reduce the impacts of hazards" would seem to encompass both elements.

From the policy perspective, if terrorism is viewed as just one of multitude of hazards that must be considered by a jurisdiction during risk assessment, then it integrates well into an all-hazards model. One assesses the risk, identifying potential targets and possible scenarios. One mitigates the risk through programs such as intelligence assessment (non-structural mitigation) or through target hardening (structural mitigation). One prepares for potential scenarios by developing plans, training personnel, and acquiring equipment. One responds if the event occurs and recovery can be done under a single jurisdictional recovery plan. If, on the other hand, one sees terrorism as something outside the all-hazards model, then one must engage in a planning process that is duplicative and runs the risk of being confusing at the time of the event. There is no question that the CEM model is applicable to terrorism.

This having been said, the success of the model has been mixed and there are some problems with it. Because of the way emergency management evolved from the Civil Defense programs, the emphasis has always been on the preparedness and response phases. As was discussed in a previous chapter, the development of the CEM concept created problems for emergency managers who had been primarily tactical and operational planners. Strategic concepts like mitigation were given lip service but were rarely adequately addressed in emergency management programs.

The available data from the National Emergency Management Baseline Capability Assessment (NEMB-CAP) supports this thesis. The study to date has found only 26 percent of the surveyed states conformant for recovery planning, 23 percent conformant with mitigation planning, and 14 percent conformant for continuity planning. This is in stark contrast to the 51 percent conformance for response planning. These figures are not a reflection of the validity of the CEM model but rather suggest that emergency managers are not adequately addressing the strategic concepts inherent in the model.

This failure can be attributed in part to the evolution of the emergency management position discussed in Chapter 3. However, it may also be the result of fundamental misunderstandings about the model itself. Since the model is based on a cycle where the phases flow in a sequential and logical manner, this has been taken by many to suggest that this is how operations progress. One finds, for example, numerous state and local emergency operations plans that are organized by the CEM phases and that attempt to clearly delineate what tasks occur in each phase. Although they do not incorporate the CEM phases, Homeland Security planning documents such as the Universal Task List and Target Capabilities List are organized in a similar fashion around the phases of prevention, protection, response, and recovery. In reality, plans are implemented virtually simultaneously with tasks related to response, continuity, recovery, and even mitigation being implemented at the same time.

This suggests that using the CEM model as the basis for emergency operations planning has limited validity. If, however, one considers the model as a strategic concept, the dynamic changes. The CEM model can be used to denote four interrelated strategies for mitigation, response, recovery, and preparedness that together define the community's response to crisis. The relationship among these strategies is a product of how the community manages risk.

RISK MANAGEMENT

Risk is managed in four basic ways. The organization can choose to simply avoid a risk by not taking an action that carries the potential for liability. The organization can transfer the risk, usually through the use of insurance. The organization can mitigate the risk through structural or non-structural methods of risk reduction. Finally, the organization can choose to retain the risk, either by self-insuring, ignoring the risk, or developing the capacity to respond to the impacts if the risk occurs. The combination of these four elements will vary from organization to organization depending on risk tolerance, vulnerability, and available resources.

A good starting point for developing a risk management strategy is an insurance review. Even where a community is self-insured, there is usually some reliance on risk transfer through third party insurance. In addition, there are costs associated with disasters that may not have been factored into the original loss estimation. In this age of tight budgets, self-insured organizations tend to view self-insurance as a statement of risk tolerance rather than actually creating cash reserves to cover extraordinary costs. This risk tolerance may be acceptable for day-to-day emergencies but becomes intolerable in a major disaster where costs become excessive and income streams are severely reduced.

Where insurance coverage exists, the full extent of the coverage is not always understood and many policyholders tend to make assumptions about what is covered and what is not. For example, does the policy cover full replacement of a demolished building or only the cost to the damaged portion? If repair of a damaged structure triggers a mandatory upgrade to an improved building code, is the added cost covered? If a facility is a crime scene and access is prevented for several weeks, is the cost of alternate operating facilities recoverable?

To the emergency manager, the insurance review is necessary for two reasons. First, recoverable expenses are critical to recovery and the restoration of the community to some semblance of normalcy. The major source of community recovery funding is the insurance industry. Second, at some point, insurance requirements and emergency plans will intersect. This means that considering actions from a claims perspective may have a bearing on response issues. For example, the training responders receive in documenting damage could be oriented toward helping to fulfill both the need for initial damage assessment data and for capturing data for later insurance claims. A decision to clear an essential roadway of debris may create issues for homeowners with their insurance company—it is not uncommon for insurance carriers to reimburse the cost of clearing debris caused by the disaster but to refuse coverage for removing debris created during road clearance by the jurisdiction. Decisions to dispose of damaged equipment or to demolish dangerous structures may create future issues during the claims process if the damage has not been fully documented in a manner considered sufficient by the insurance company.

Insurance language can be confusing, as terms may have different meanings to insurance professionals and emergency managers. Everyday terms such as useable, debris, salvage, replace, and recover have very different and specific meanings in the insurance industry. Using such common terms in the wrong context may have an impact on the settlement of a claim. Consequently, an important addition to the emergency planning team is the community risk manager.

There would seem to be a natural alliance between the risk manager and the emergency manager. The risk manager seeks to protect the jurisdiction against preventable losses. The primary tool of the risk manager is the risk assessment, using much the same methodology of hazard identification and risk analysis employed by the emergency manager.

MITIGATION STRATEGY

During the risk analysis, it is possible to identify opportunities that could either eliminate the organization's vulnerability to a hazard or substantially reduce the impact of the hazard. These measures are referred to as mitigation and are distinguished from preparedness in that they are generally long-term measures focused on reducing or eliminating the impacts of a hazard as opposed to enhancing the capacity to respond. Because it seeks to be proactive, mitigation offers the potential for significant savings both in the elimination or reduction of damages and in the reduction of response capacity. Mitigation can also have an impact on insurance premiums, which are risk based, and can result in significantly lowered costs. For these reasons, James Lee Witt, former Director of FEMA, was frequently quoted as saying that "mitigation is the cornerstone of emergency management."

Yet mitigation plays only a limited role in most communities and is generally applied post-disaster to prevent losses if a similar event should occur. There are several reasons for this. Mitigation is a strategy and, as such, requires consideration from a strategic viewpoint. Most emergency planners have mainly tactical expertise and lack both the skills and the organizational standing to influence strategic decisions. Further, the development of mitigation strategies requires a different set of stakeholders than those involved in the development of emergency plans. Many of these disciplines are foreign to emergency planners and require specialized expertise. Finally, mitigation planning, even more so than emergency planning, requires the involvement of community, political, and popular leadership.

Most jurisdictions have treated mitigation as they have other elements of the emergency management program—they have assigned it to a single department, such as the planning department, developed a document based on a template, and failed to provide any significant funding. The same is true of the federal government. Federal mitigation funding has traditionally been provided post-disaster under the Hazard Mitigation Grant Program (section 404 of the Stafford Act). This funding is a percentage of the total grants awarded by the federal government under a presidential declaration of disaster. During the Clinton administration, this percentage was increased to a high of 25 percent, but under the revisions to the Stafford Act promulgated by the Disaster Mitigation Act of 2000, the Hazard Mitigation Grant Program was limited to 7.5 percent, providing the state has an approved standard mitigation plan. If the state has an enhanced mitigation plan, the president may increase this amount up to 20 percent.

Realizing the post-disaster mitigation is the policy equivalent of locking the barn door after the horse has run away, DMA 2000 established an innovative pre-disaster mitigation program. Although limited in scope (grants are small and awarded competitively), the program acknowledges for the first time that pre-disaster mitigation should be preferred to post-disaster mitigation. Here again, though, government support has been lacking. Under the Bush administration, the program has been cut from a high of $255 million in FY 2005 to $50 million in FY 2006.

Social science research indicates that for many jurisdictions there is very little correlation between objective risk, perceived risk, and mitigation efforts. Researchers found that communities were aware of potential hazards, but for a variety of reasons opted not to take action. Mitigation is never simple and may generate resistance on the basis of the perceived violation of property rights or hindering development. It may also have an impact on those of lower socio-economic status and be perceived as favoring the rich over the poor. There are also examples where communities have adapted to recurring events in such a way that mitigation is of little interest and would be considered disruptive to their society. In addition, the lack of leadership from the federal government, low priority given to mitigation by state and local governments, and absence of clear financial incentives for mitigation creates an environment that is not conducive to mitigation planning.

One key point that emerges from the social science literature is the importance of a "champion" for mitigation. Even where disaster creates

CASE STUDY: ST. CHARLES COUNTY, MO—CONFLICTS IN MITIGATION

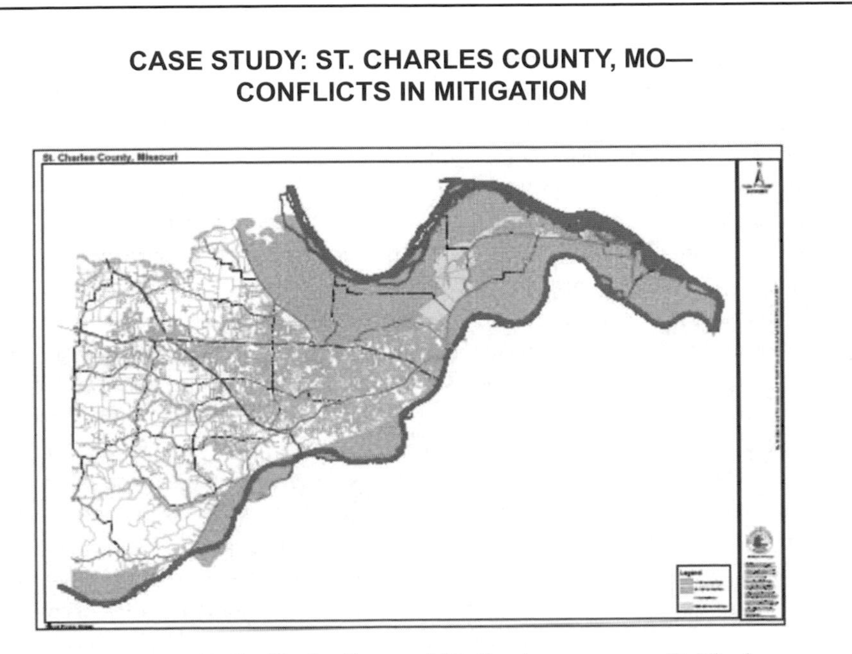

Case Figure 6.1 St. Charles County, MO, flood prone areas—St. Charles County government.

St. Charles County, MO holds the record for repetitive claims under the National Flood Insurance Program. The county of 300,000 sits at the confluence of the Mississippi and Missouri Rivers and almost half the county is located in a floodplain (see Case Figure 6.1). The risk of flooding has been exacerbated by farming that reduced the amount of wetlands and by a federal levee program demanded by the farming community. There is also a local attitude that floods are natural and not subject to human intervention.

The Midwest Floods of 1993 hit St. Charles hard. Over 2,100 homes were condemned due to flood damage. Following the disaster, St. Charles County agreed to participate in the Missouri Buyout Program, a mitigation plan backed by FEMA to reduce risk by purchasing properties in high-risk zones. From 1993 to 1995, the county used $5.78 million in Hazard Mitigation Grant Program funding from FEMA and $8.8 million from the Community Development Block Grant Program to acquire 1,159 properties. The acquisition of these repetitive damage properties prevented losses in 1995 and damage in a 2002 flood was so insignificant that it did not warrant a presidential declaration of disaster, even though other counties in the area suffered significant damage. FEMA rightly points to St. Charles County as a mitigation success story.

(continued on next page)

(continued from previous page)

However, there is a darker side to the story. Major development in the 1950s had seen the development of three major trailer parks and the construction of inexpensive housing in the main area of flooding. The advent of the National Flood Insurance in 1968 coupled with the county's unwillingness to strictly enforce its provisions encouraged those with low incomes to take advantage of the cheap housing. Since existing mobile home parks were the only housing allowed to be reconstructed below the 100-year flood mark and still retain insurance coverage, the majority of these residents lived in mobile homes. Attempts in 1986 to make these mobile homes safer by elevating them above the 100-year flood mark were bitterly resisted by both the county and the National Manufactured Housing Federation, as were requests by the residents to relax county zoning ordinances to allow relocation of the mobile home parks.

When the Missouri Buyout Program was implemented, the burden of mitigation fell directly on the poor and disadvantaged. The owners of the mobile home parks were compensated for their land under the program. The renters, however, received nothing, as mobile homes were considered personal property and not real estate. Plans to build a new subdivision of affordable housing collapsed after resistance from local residents. The only option for many was to leave St. Charles. In one group of 2,800 homeless families, 90 percent were estimated to have relocated to other counties.

St. Charles County is an example of the complexity of mitigation. It shows how even well intentioned programs like the National Flood Insurance Program can have unanticipated results. It demonstrates that simply reducing risk is not always sufficient nor is it as easy as it would appear—there are always tradeoffs in any policy decision. Effective mitigation strategy must include not only risk reduction methodology, but also consideration of the effects of implementing that methodology.

a window of opportunity, hazard reduction is unlikely to occur without the involvement of individuals or organizations prepared to push for the adoption of mitigation strategies. This suggests that it is possible for the emergency manager to create the impetus for mitigation.

The key to developing this impetus is an understanding that mitigation is not simply a technological solution for loss reduction, but is socially structured. This means that disasters are not the product of "Acts of God" outside human control but of the interaction of the hazard and the community's vulnerability. Like vulnerability, mitigation is influenced by the community. In other words, the decision to mitigate and the strategies selected will be determined by the sociocultural factors of the community. The assumption that one need only provide an effective technical proposal for mitigation to occur is a false one. Even strategies that have been proven effective in other situations may fail to gain the acceptance of a community. Thus, proposed mitigation strategies must be understood within their social context and must be developed in a social environment that assists in their acceptance. In the words of Dr. Kathleen Tierney, "Mitigation strategies typically stand or fall on their political, economic, and sociocultural feasibility—not on their technical feasibility." (Tierney 1993).

When seeking community support, one invariably runs up against the well-meaning person or organization that insists on the adoption of all potential mitigation measures. While it has been demonstrated that not all mitigation measures will be acceptable, a further limiting factor is the issue of cost-effectiveness. There are those who feel that cost should not be an issue in hazard reduction, but the simple fact is that cost-benefit analysis is an acceptable method of determining the appropriateness of mitigation measures. As was pointed out, if there is no perceived value to the mitigation measure, then it will not be acceptable to the community, particularly when mitigation resources are limited.

Based on the social science research, it becomes apparent that the common approach of developing a mitigation plan in isolation to meet DMA 2000 requirements does little to ensure effective community hazard reduction. One can also begin to draw some conclusions about what will be necessary for successful mitigation:

- A champion, or champions, to spearhead the mitigation effort must be identified. While this can be the emergency manager, it is more likely that the emergency manager will serve as a catalyst to identify and motivate community organizations and politicians.
- Mitigation measures must be cost-effective and seen as adding value to the community.

- Mitigation measures must be acceptable within the sociocultural and socioeconomic norms of the community.
- Mitigation measures must be debated and adopted under an open process that stimulates public involvement and acceptance.

The implication of these elements of success is that mitigation planning must be inclusive and broad based. This is supported by 44 Code of Federal Regulations Emergency Management Assistance, the document that contains the rules under which federal disaster assistance programs are administered. 44CFR requires that the planning process provide for public comment, involvement of neighboring communities, businesses and regulatory agencies, and the review and incorporation as appropriate of existing technical reports and studies.

Mitigation strategy is ultimately aimed at reducing or eliminating the potential impacts of hazards. While the Stafford Act addresses natural hazards, it would be imprudent for a jurisdiction to pursue a mitigation strategy that did not include provisions for human-caused hazards as well. Such a multi-hazard strategy is in keeping with the tenets of comprehensive emergency management and allows the community to leverage both Stafford Act and Homeland Security funds under a single comprehensive program of risk reduction.

Policy makers should consider both pre- and post-disaster mitigation strategies. Pre-disaster strategies can be further divided into actions that can be taken immediately at little or no cost and those that can be accomplished over time with anticipated or existing funding. Pre-disaster mitigation does not necessarily have to be structural and programs aimed at reducing non-structural hazards (e.g. earthquake bracing, typhoon clips, etc.) can be relatively inexpensive. Mitigation strategies can also have multiple uses: a public warning system could potentially be used during civic events.

Mitigation strategy should also identify actions that can be implemented following a disaster as part of a holistic recovery strategy to improve community quality of life. Such strategies should be a logical extension of pre-disaster mitigation and include projects that are desirable but either lack funding or the political will to implement.

RECOVERY STRATEGY

Recovery is the transition from disaster response back to an acceptable state of normalcy. It would be wrong to say that recovery returns the community to normal for two reasons: 1) the widespread destruction and social disruption caused by major disasters produce profound changes that prevent the community from fully returning to its pre-disaster condition; and 2) disasters offer opportunities for changes in the community that can create a new "normal." Like mitigation, recovery planning is an extremely low priority for local governments. Part of the reason for this is the perception that recovery is "the last thing we do" and therefore accorded a lower priority than response planning. Part of it stems from a failure to understand potential recovery issues.

Since recovery is the last phase in the emergency management cycle, it seems strange to consider it before one discusses emergency response. However, recovery is not a single process but a series of complex processes, many occurring simultaneously. Without a strategy to serve as a guide, these processes can lead to conflicting priorities and extreme political and social upheaval. Further, without a well-defined strategy, decisions made in the early phases of response foreclose options for recovery, severely limiting the options available to the community. Finally, the recovery period offers an opportunity to implement components of the mitigation strategy and the two must be closely linked.

Eugene Haas and his colleagues identified four overlapping periods that occur in recovery:

- An emergency period that covers the immediate aftermath of the event and is focused on coping with immediate losses
- A restoration period that covers from the end of the emergency period to the restoration of major services
- A replacement reconstruction period that results in the rebuilding of capital stock and the return of social and economic activities to pre-disaster levels
- A developmental reconstruction period that provides for major reconstruction and future growth

Like many of the previous disaster models discussed, these four periods should not be seen as distinct. In actual practice, there is considerable overlap and it is not uncommon for different segments of the community to be in different periods owing to differences in economic resources. Pre-disaster recovery planning can, to a certain extent, shorten the length of these periods and speed recovery.

As one examines the four periods of recovery, it is possible to separate them into short-term and long-term activities. Short-term activities that lead to the restoration of basic services are generally tactical-level activities, such as debris clearance and the repair of basic infrastructure. The focus is on restoration of community services to near-normal levels. A major component of this restoration is the use of federal disaster assistance programs for individual and public assistance. Short-term recovery strategies should be oriented toward a detailed damage assessment (e.g. building inspections, structural assessments of bridges and roads, etc.) and establishing an interface with federal programs.

Long-term recovery, however, is strategic in nature because of its potential impact on the community's future. Like mitigation, long-term recovery strategy requires community acceptance to be effective. That this acceptance must be obtained before the disaster is obvious when one considers the potential barriers to developing and implementing a strategy post-disaster. In contrast to the altruism found during the response period, the recovery period is characterized by conflicting priorities, concern over perceived inequities in disaster relief and a tendency to establish blame for inadequate response. The overwhelming community emphasis is on a rapid return to normalcy. This means that barriers to quick rebuilding, such as a public process to agree on mitigation methods, are considered unacceptable. Indeed, there is usually a push by the community for a relaxation of building codes and restrictions on development to spur reconstruction, actions counterintuitive to increasing community resilience.

Communities tend to want to replicate pre-disaster conditions, so the focus of recovery tends to be on returning the community to the way things were before. One sees this dynamic at work in the aftermath of Hurricane Katrina. When the Bring Back New Orleans Commission released its initial recommendations in January 2006, it recommended a four-month moratorium on rebuilding to determine what neighborhoods would have enough returning residents to warrant rebuilding. There was concern that some areas would be too thinly populated to be sustainable. The recommendations were immediately met by angry protests from citizens, condemnation by groups such as the NAACP, and attempts by permitting officials to subvert the plan by issuing special construction permits to anyone who asked for one. Despite the opportunity to reduce risk in those areas where flooding was the worst, returning residents are demanding a rebuilding of their neighborhoods in hopes of reestablishing the communities that existed prior to the hurricane.

Local government, seeking to avoid criticism for slow recovery, will generally accede to the community's demands for quick action. The emphasis is on acquiring tools and techniques to quickly restore the status quo and expedite federal payments. The time required to form a recovery task force and to develop long-range strategy is overwhelmed by the rapid pace of reconstruction. An example of this problem is given in the case study involving the aftermath of the 1906 San Francisco earthquake and fire. In this case, a pre-disaster proposal for a new city layout was considered as a blue print for reconstruction but ultimately rejected because of the speed of rebuilding.

The paradox of recovery is that, while the period immediately following the disaster is clearly not the time to begin development of a reconstruction plan, it is the time when the public is prepared to show an interest in recovery and mitigation issues. The first 30 days following a disaster is the critical window of opportunity to establish the mechanism for managing and guiding the recovery. The lesson is clear; recovery strategy development must be done before the disaster if one wishes to take advantage of this window of opportunity.

Effective recovery planning requires four main components:

1. Formation of a broad-based task force that represents the community as a whole
2. Development of a strategy document that represents a community vision and consensus
3. Development of an operational plan to cover short-term recovery
4. Passage of ordinances providing authorities for recovery activities

CASE STUDY: SAN FRANCISCO 1906—MISSED OPPORTUNITIES IN RECOVERY

Case Figure 6.2 San Francisco Financial District following the earthquake and fire. Courtesy of the Virtual Museum of the City of San Francisco.

The devastation of the earthquake and fire that occurred in San Francisco on April 18, 1906 was certainly on a massive scale: the area burned by the fire was twice that of Chicago Fire of 1871 and six times that of the Great Fire of London in 1666. Three quarters of the developed area of the City were destroyed along with much of the civic infrastructure. Over half the City's population of 410,000 was displaced. Property loss was estimated at between $500 million and $1 billion (in 1906 dollars). The massive relief effort that followed the earthquake and fire lasted over three years.

With the city virtually leveled, it was imperative that reconstruction begin immediately. By a strange twist of fate, San Francisco already had a plan for a complete reconstruction of the city, the Burnham Plan. As mayor, James Phelan had attended the Chicago World's Fair in 1893 and been very impressed by the concepts of architect James Burnham. Phelan brought Burnham to San Francisco to develop a radical redesign of San Francisco. Based on the design of European capitals, Burnham's plan envisioned broad boulevards, open spaces, and expansive parks. A grand civic center was to be located at Van Ness and Market Streets and nine broad boulevards would radiate outward, connected by concentric streets. Burnham's plans were completed just a few days before the earthquake. Now, as chair of the powerful Finance Committee, Phelan pushed for adoption of the Burnham Plan.

(continued on next page)

(continued from previous page)

The Burnham Plan had several flaws. First, to accomplish the plan would require property owners to sell existing parcels and acquire new ones. This was particularly true in Chinatown, where there was a push to evict the entire Chinese population from prime real estate and relocate them in less desirable parts of town. Secondly, the Burnham Plan did not take into account the economic heart of the city: there was no consideration for the commercial, industrial, and waterfront areas in the plan. Nevertheless, Phelan and his colleagues in the Association for the Improvement and Adornment of San Francisco pushed hard for the adoption of the Burnham Plan.

Arrayed against Phelan and his powerful friends was the tendency of disaster victims to attempt to return as closely as possible to their former situation. As was noted in Chapter 2, disaster victims are resilient and will begin recovery on their own. In most cases, this return to normalcy involves replicating their previous situation. This is why post-disaster mitigation is difficult to implement. The citizens of San Francisco simply were not going to wait until the details of the Burnham Plan could be worked out. Similar issues arose after Hurricane Katrina, where citizens insisted on rebuilding their homes in opposition to a plan to relocate neighborhoods to safer areas.

Phelan and his supporters attempted to pass a state constitutional amendment that would allow the city to acquire land to implement the Burnham Plan by trading property. The amendment was defeated in a statewide election. Opposition also came from the downtown business leaders who would not countenance any delay in rebuilding. The loss of city records, including property titles, meant that implementing the Burnham Plan could take some time. This lengthy delay would have an obvious detrimental effect on business resumption and on the tax base. Ultimately, private property rights won out over public good and the Burnham Plan was shelved.

The Burnham Plan could have transformed the city of San Francisco into a grand city on the order of Paris or Berlin. Given time and political will, it is possible that the details of the plan could have been worked out and the plan implemented. However, in the aftermath of catastrophe, there was simply no time to identify the legal and social implications of the plan and to work out compromises. Further, the plan did not enjoy the full support of the community. It was favored by the wealthy upper class but was deemed impractical by hardheaded businessmen. The Burnham Plan is a reminder that recovery planning, no matter how visionary, must be timely and supported by the community.

In forming the recovery task force, one must confront the issue of governance. Is the task force responsible solely for the purpose of developing the recovery strategy or will it actually guide the process as well? If the latter, what is its relationship to the group handling response? If it is solely for planning, who will guide the recovery effort? The answers to these questions will help to determine the membership of the core group and help to identify additional stakeholders in the planning process.

As has been mentioned, recovery involves multiple processes. In may be beneficial to consider dividing the task force into smaller committees that focus on specific issues, such as:

- Social Recovery—issues that affect the community directly, such as the restoration of housing stock, the reopening of schools, resumption of social services, etc.
- Infrastructure Recovery—the physical restoration and reconstruction of community infrastructure (e.g. public works, civic buildings, road construction, etc.).
- Economic Recovery—restoration of economic vitality by re-establishing businesses, attracting tourism, stimulating investment, etc.
- Environmental Recovery—issues related to control of cleanup of hazardous waste, landfill capacity, debris disposal, etc.

The development of a strategic recovery plan presents the opportunity to go beyond the status quo and to seek advantages inherent in a crisis. The Natural Hazards Research and Applications Information Center has produced a blueprint for what has been termed "holistic recovery," where the focus is not on a return to pre-disaster conditions but on increasing community sustainability. "Sustainability" in this context means "development that meets the needs of the present without compromising the ability of future generations to meet their own needs." Holistic recovery is based on six principles of sustainability:

1. Maintain and, if possible, enhance quality of life
2. Enhance local economic viability
3. Promote social and intergenerational equity

4. Maintain and, if possible, enhance the quality of the environment
5. Incorporate disaster resilience and mitigation into decisions and actions
6. Use a consensus-building, participatory process when making decisions

The six principles of holistic recovery summarize the intent espoused but seldom achieved by emergency planners. In essence, they offer a methodology to not only deal with the immediate consequences of a disaster but to use the disaster as a mechanism to increase resistance to future disasters. They also offer the opportunity to link the recovery and mitigation strategies to overall community strategies, increasing the involvement of community groups and the potential for community acceptance. By emphasizing that recovery and mitigation strategies are congruent with community goals, the emergency manager can demonstrate clearly the value of the strategies.

Once one knows the end result, the intermediate steps are easily derived. To implement the strategy, there may be a need to craft recovery specific legislation to institutionalize the planning process, establish the governance structure, and address specific planning issues. Figure 6.2 is a table of contents for a model ordinance that suggests some of the issues that might be considered. Passing an ordinance validates the work of the recovery task force and codifies the authorities that will be needed during recovery.

Short-term recovery is also informed by the strategy. Since long-range recovery goals are known, responders can make decisions consistent with these goals and lay the foundation for later implementation of the full strategy. Thus, the operational recovery annex and the long-range recovery plan become the implementing documents for the recovery strategy.

RESPONSE STRATEGY

If done properly, the hazard analysis develops a summary of the risks facing a community in terms of vulnerabilities and hazards. The mitigation strategy limits this field by either eliminating hazards or reducing their potential impact. The recovery strategy defines the end

Section 1. Authority
Section 2. Purpose
Section 3. Definitions
Section 4. Recovery Organization
4.1 Powers and Duties
4.2 Recovery Task Force
4.3 Operations and Meetings 4.4 Succession
4.5 Organization
4.6 Relation to Emergency Management Organization
Section 5. Recovery Plan
5.1 Recovery Plan Content
5.2 Coordination of Recovery Plan with FEMA and Other Agencies
5.3 Recovery Plan Adoption
5.4 Recovery Plan Implementation
5.5 Recovery Plan Training and Exercises
5.6 Recovery Plan Consultation with Citizens
5.7 Recovery Plan Amendments
5.8 Recovery Plan Coordination with Related (City, County) Plans
Section 6. General Provisions
6.1 Powers and Procedures
6.2 Post-Disaster Operations
6.3 Coordination with FEMA and Other Agencies
6.4 Consultation with Citizens
Section 7. Temporary Regulations
7.1 Duration
7.2 Damage Assessment
7.3 Development Moratorium
7.4 Debris Clearance
7.5 One-Stop Center for Permit Expediting
7.6 Temporary Use Permits
7,7 Temporary Repair Permits
7.8 Deferral of Fees for Reconstruction Permits
7.9 Nonconforming Buildings and Uses
Section 8. Demolition of Damaged Historic Buildings
8.1 Condemnation and Demolition
8.2 Notice of Condemnation
8.3 Request to FEMA to Demolish
8.4 Historic Building Demolitions Review
Section 9. Temporary and Permanent Housing
Section 10. Hazard Mitigation Program
10.1 Safety Element
10.2 Short-Term Action Program
10.3 Post-Disaster Actions
10.4 New Information
Section 11. Recovery and Reconstruction Strategy
11.1 Functions
11.2 Review
Section 12. Severability

Figure 6.2 Model Recovery and Reconstruction Ordinance. From *Planning for Post-disaster Recovery and Reconstruction,* American Planning Association.

state—the point at which the community achieves an acceptable state of normalcy. Together these two strategies bracket and inform a third: the strategy that the community will use to respond to the immediate impact of the event, maintain continuity of function, and initiate short-term recovery.

Few communities actually develop a response strategy as opposed to an emergency plan. This is most likely because much response planning is done by rote. That is, communities use templates for emergency plans such as that provided by FEMA's State and Local Guide 101 or one provided by a consultant or a software program. There are valid reasons for using such an approach—it ensures compliance and conformance with federal guidelines and ensures that all critical areas are addressed. However, many of these planning guides are geared toward initial response and do not consider other plans that might be implemented simultaneously. For example, SLG 101 does not address recovery, continuity of operations, or post-disaster mitigation.

Emergency plans are by nature tactical documents, not strategic ones. They tend to focus on short-term issues related to response rather than on strategy and policy. This is their purpose and the process involved in developing emergency plans can serve to raise issues of strategy and policy. However, the development and agreement on strategy beforehand simplifies the development of the various plans needed to respond to crisis. Divorcing plan development from the strategic planning process results in plans that are disconnected from each other and may lead to conflicts among the plans. It is at the strategic level that interrelationships among plans become most clear.

As has been noted in previous chapters, response, continuity, and recovery do not occur sequentially. Separate plans are implemented simultaneously or within a very short time frame. If these plans are developed in a vacuum, this simultaneous implementation may result in conflicts over management of the crisis and a competition for scarce resources. It is critical, therefore, that concepts for response be worked out prior to beginning to develop plans and that significant policy issues be addressed.

A critical part of response strategy is the development of a governance structure and the fixing of responsibility for various response functions. Response involves the virtually simultaneous implementation of emergency, continuity, and recovery plans. A major strategic decision is whether these plans are coordinated from a single operations center by a single management team or from separate operations centers headed by different managers. If the latter, how will conflicts be resolved? Who is the final arbitration authority? Will there be a single incident action plan? If so, who develops it? Another issue of governance involves the transition of operational control. Immediate response is relatively short-lived and the jurisdiction moves very quickly into sustained operations, focusing on recovery issues. How will this transition be handled? At what point does the central point of coordination move from an emergency operations center to something more long term?

In addition to these types of governance issues, conflicts frequently arise because priorities are not defined. Life safety is always the first priority in response, but as was noted in a previous chapter, one of the qualitative differences between disasters and emergencies is a change in speed of response and in standards of care. Hence, the operational dynamic changes and while life safety is always paramount, there will be decisions made that will adversely impact victims, such as the use of triage for prioritizing treatment to the injured and the imposition of priorities for distribution of limited supplies. It is this changing dynamic that needs to be addressed as the response strategy is being developed. How much compromise is the community willing to tolerate to ensure its own survival?

An example of this dynamic occurred during the 1989 Loma Prieta earthquake in San Francisco. The Marina District, an affluent area built on filled ground, was heavily damaged and suffered significant structural collapse. As a safety measure, authorities evacuated the area and prohibited reentry by the occupants. Plans were made to begin immediate demolition of unsafe structures. From the city's perspective, the priority was life safety. The occupants, however, were horrified that they could not reenter and salvage their personal possessions before their homes were destroyed. Ultimately, the city authorities yielded to the demands of the community and allowed escorted reentry but for only 15 minutes per occupant. This solution, as would be expected, pleased no one. Identifying this need, developing a reentry strategy, and assigning responsibility for planning and implementation could have minimized this conflict.

The purpose of the response strategy is to articulate a concept for tactical response, to identify potential policy issues that may arise during the course of operations, and to fix responsibility for planning and implementation. One begins this process by determining the functions that will be required in a disaster. Figure 6.3 is a list of functions developed by EMAP that represents the minimum areas that should be addressed in an emergency operations plan. At this stage, details are not important—the planners are seeking to develop a concept for each function and to identify the agencies that will do the detailed planning.

As an example of this process, consider the function of evacuating the community. Evacuation is surprisingly difficult because such operations almost always involve a transfer of responsibility. When a com-

Direction/control and coordination;
Information and planning;
Detection and monitoring;
Alert and notification;
Warning;
Communications;
Emergency public information;
Resource management;
Evacuation;
Mass care;
Sheltering;
Needs and damage assessment;
Military support;
Donated goods;
Voluntary organizations;
Law enforcement;
Fire protection;
Search and rescue;
Public health, medical, and mortuary services;
Agriculture;
Animal control/management;
Food, water and commodities distribution;
Transportation resources;
Energy and utilities services;
Public works and engineering services; and
Hazardous materials.

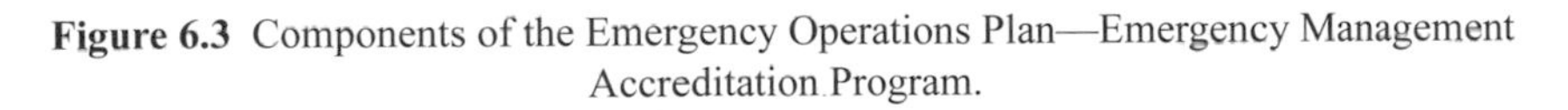

Figure 6.3 Components of the Emergency Operations Plan—Emergency Management Accreditation Program.

munity is evacuated, the evacuees become the concern of the other communities through which they must pass and of the community that will ultimately receive them. Even the evacuation of a single facility will have an impact on the surrounding area. Consequently, if the community's strategy includes the need to partially or fully evacuate, then a considerable number of policy issues arise from that decision:

- Will evacuation be full or partial? Traditionally, evacuation planning was based on natural hazards. The Cold War requirement for evacuation planning under the Crisis Relocation Program was strongly resisted by local governments as unnecessary. However, the potential use of weapons of mass destruction, particularly dirty bomb scenarios, has caused jurisdictions to rethink their strategies.
- Will spontaneous evacuation be supported? There are any number of conditions that may cause people to self-evacuate. Such self-evacuation can create significant problems as the control meas-

ures and supporting infrastructure that are in place during mandatory evacuations may not be present. On the other hand, deploying these resources may draw scarce resources from other tasks.

- Under what conditions will evacuation be mandated? Evacuation is costly and can expose evacuees to considerable risk. It is not a decision to be made lightly. Further, as has been demonstrated in numerous hurricanes along the Gulf Coast and the East Coast, frequent evacuations lead to a certain complacency and unwillingness to evacuate. There is also evidence of unwillingness on the part of officials to order evacuation, even when pre-identified conditions have been met.
- What authority does the leadership of the community have to order an evacuation? Can field personnel, such as a fire officer, order evacuation? Is evacuation mandatory and can citizens be forcibly removed? Can the media be restricted from entering the evacuated area?
- How will barriers to evacuation be overcome? Hurricane Katrina demonstrated that there are segments of the population that cannot or will not self-evacuate. Many of these were people without the means to do so or people with disabilities that prevented them and their caregivers from leaving. Statistics indicate that this can be a substantial part of the population: the 2000 census indicated that between 21 and 25 percent of the population of New Orleans, Biloxi, and Mobile described themselves as disabled. In a Kaiser Foundation poll taken in Houston of people evacuated from New Orleans, 22 percent of respondents stated that they were physically unable to leave and a further 23 percent stated that they had to care for someone unable to leave.

Closely associated with evacuation planning is the need for sheltering. Many emergency plans treat sheltering as something that happens almost immediately. In actual practice, it may take several days to inspect, staff, and equip shelters to their full capacity. Congregate shelters are also a short-term measure; the Red Cross expects that they will be maintained for no more than 30 days. They buy time for either the individual or the jurisdiction to find interim housing. Sheltering must therefore be viewed as a continuum ranging from immediate temporary shelters followed by transition to short term housing and eventually leading to permanent housing. A shelter strategy must address issues such as:

- What constitutes an acceptable shelter? For example, one jurisdiction designates as official only those shelters that meet ADA requirements but maintains a second list of shelters that do not currently meet the requirements but could be used with some modifications during a disaster.
- Will persons with disabilities be housed at all shelters or will there be special shelters established to accommodate special needs? If so, what is the definition of "special needs" and will the shelter accommodate caregivers?
- How long are shelters expected to be open? The Red Cross initially had responsibility for long term sheltering but their policy is now to maintain shelters for about 30 days at which point FEMA is expected to assist in the transition to short-term housing.
- Will existing buildings be used for sheltering or will camps be established?
- How will short-term housing be provided? Will semi-permanent camps be established as was done in San Francisco in 1906 and Kobe, Japan in 1995? Or will victims be removed from the disaster area and housed in hotels, as was done after Hurricane Katrina in 2005?

This discussion so far has considered the emergency operations plan and is on familiar ground for most emergency managers. However, the intent here is to consider strategy and to do that, one must stretch the definition of response to "those actions and activities that occur prior to, during, or in the immediate aftermath of a disaster." This means that instead of the focus on traditional response functions that characterize emergency management planning, one must ask the question from a broad perspective. For example, planners must take into account Recovery Time Objectives (RTOs) that must be met to sustain critical government functions. This means that response strategy must encompass not only the emergency operations plan but Continuity of Operations Planning (COOP) as well. Further, it must also consider the short-term recovery planning developed as part of the recovery strategy.

Continuity planning is a new concept to many emergency managers but it has actually been around for a long time. The idea of ensuring continuous government operations dates back to at least 1982 with the issuance of National Security Decision Directive 55 Enduring National Leadership dated September 14, 1982. The most recent iteration, Presidential Decision Directive 67, Enduring Constitutional Government and Continuity of Government Operations, dated October 21, 1998, established Continuity of Operations (COOP) as a unifying concept to ensure the continuance of essential government functions. In 2003, the Federal Emergency Management Agency's Office of National Security Coordination was designated as the lead agency for the federal executive branch COOP program. In addition to FEMA providing guidance, each agency was responsible for appointing a senior official to serve as program manager for COOP planning, programming, and budgeting and was expected to have COOP plans in place by October 1999.

In February 2004, the General Accounting Office (GAO) issued a report entitled *Continuity Of Operations: Improved Planning Needed to Ensure Delivery of Essential Government Services* that severely criticized government agencies' adherence to the FEMA COOP guidance promulgated in Federal Preparedness Circular 65 Federal Executive Branch Continuity of Operations (COOP). Of 34 agencies surveyed, the GAO found none to be in full compliance with FPC 65. The GAO also faulted FEMA for failing to include criteria for essential functions in its guidance and for failing to review essential functions when assessing COOP plans.

FPC 65 was reissued on June 15, 2004 to meet the GAO's directive to provide more defined criteria for COOP. As part of these criteria, FPC 65 establishes the requirement for access to a local area network (LAN) and vital files, records, and databases within 12 hours of COOP plan activation. In addition, the COOP plan must provide for sustained operations for up to 30 days until normal business activities can be reconstituted.

In addition to the criteria for a 12-hour RTO and sustained capability for 30 days, FPC 65 identifies eleven elements that are necessary for a viable COOP capability:

1. Plans and procedures
2. Essential functions
3. Delegations of authority
4. Orders of succession
5. Alternate operating facilities
6. Interoperable communications
7. Vital Records and databases
8. Human capital
9. Test, training, and exercises
10. Devolution of control and direction
11. Reconstitution

Many of the elements on the FPC 65 list are already part of the emergency management program and plans. More importantly, COOP aims at preserving the baseline capability necessary for tactical and operational response. Given the level of devastation in recent disasters, one can no longer just assume that government agencies will be able to fulfill their disaster missions in the absence of continuity plans. As continuity plans become more common, they will come to represent a source of competition for resources in the immediate aftermath of a disaster. Consequently, continuity issues represent a factor that must be considered in developing a response strategy.

The simultaneous implementation of strategies suggests that, instead of focusing on individual plans, planners should consider plans in a time-based continuum and assess what activities occur immediately, in the short-term, and long-range. Response strategy should work toward blending traditional response with recovery, continuity, and mitigation. This does not mean that all these strategies combine into a single plan. However, it does suggest that there should be a single unifying response strategy and that plans should be expanded beyond the limits of SLG 101 to encompass all immediate and short-term activities.

PREPAREDNESS

The three strategies of mitigation, response, and recovery provide for a continuum of disaster response that defines a beginning point, a desired end state, and a process for linking the two. As these concepts are developed, planners should be gaining an understanding of potential shortcomings that would impede their implementation. In essence, the combined strategies delineate the capacity that will be needed for successful response in accordance with the community's vision. The preparedness strategy is the mechanism by which the community builds its capacity to respond.

Preparedness strategy focuses on three main components: tactical planning, logistics management, and training. It defines the shortfall between current and desired capabilities for each component and options for eliminating this shortfall.

Tactical planning consists of the plans and procedures necessary to support the strategy. This includes the development of tactical-level plans for emergency response, continuity, recovery, and mitigation, and the supporting plans, operating procedures, checklists, etc. needed to implement them.

Logistics management looks at response from a resources perspective: what will be needed, what is currently on hand, and how will shortfalls be met? Based on the response strategy, it should be possible for planners to derive a list of resources required by the community and the timeframes during which they will be needed. The planners use this list to conduct an assessment of the resources on hand. This resource inventory accomplishes two things: it identifies shortfalls and develops a list or database that can be used to locate resources during response.

Some emergency management offices go to considerable lengths to create massive single lists of disaster resources. However, maintaining a central list can be labor intensive and these types of lists tend to go out of date rapidly. This technique is fairly common, particularly where responsibility for the success of a response operation rests with a single office. However, if one views the responsibility for successful response as an organizational, rather than an individual responsibility, then it is reasonable to approach resource management as an organizational function. This means that responsibility for maintaining inventory lists rests with those elements of the jurisdiction most likely to have a need for them. For example, the department of public works or the corporate engineering department would maintain a list of contractors with heavy lift capacity. Likewise, the department of public health would maintain a list of medical supplies or the names of contractors providing mental health services.

Once resource shortfalls have been identified, the organization needs a strategy for how these resources will be obtained. One possibility is the use of mutual aid, which will be discussed in more detail in a later chapter. A second source is the use of private or non-profit resources. For example, the response strategy may have identified the need for sheltering several thousand people. For most jurisdictions, the Red Cross usually takes on this task, so many of the resources needed for this task will be provided by that organization. The response strategy identified the need for sheltering and set the target population; the preparedness strategy identifies the resources needed and, through discussion with the Red Cross, what will be provided and what must still be located. In addition, the preparedness strategy should now include a component for a detailed shelter planning with the Red Cross and supporting agencies.

Trained personnel are no less a resource than specialized equipment. As was noted in a previous chapter, many communities approach training in a piece-meal fashion rather than considering it in relation to overall strategy. Training should be geared to meeting the demands for capability that are generated by the community's strategic vision. This helps to define the scope of the required training. Identifying training needs is similar to identifying resource needs: one considers the skills that will be needed to implement the strategies, conducts an inventory of available skills, and develops options for meeting the shortfall. Determining need also requires acknowledging that training is perishable. That is, it must be reinforced or refreshed if skills are to be kept current.

Training, much like response, has different levels that help define the scope of the training required. For most members of the organization, the only requirement is for awareness training that addresses potential risk and what is expected of the member (e.g. ICS100).

There is a smaller subset of trainees who require training on skills needed to accomplish specific functions under the various operational plans. For example, the teams that staff the emergency operations center will require detailed instruction in their assigned functions or a hazardous materials response team will need training on protective equipment. Some of these requirements will be defined by law, some by common sense, and some by operational need. Finally, there is a small group who will need training to provide leadership during a crisis, i.e. branch, unit, and team leaders, incident commanders, EOC directors, etc.

CONCLUSION

As was discussed in a previous chapter, the strategic plan is the mechanism by which the community translates strategy into performance objectives. This translation is accomplished by developing a series of strategies based on the community's risk and values and the baseline strategic information discussed in Chapter 4. This combination represents the ideal—a vision of how the community would like to build resilience to risks. The strategic planning process ultimately produces a preparedness strategy that articulates the difference between that ideal and the current state of the emergency management program. From this preparedness strategy, one can then derive the performance objectives in the strategic plan. Figure 6.4 diagrams this process.

The following example illustrates how the process works:

As part of its hazard assessment, a jurisdiction identifies a portion of its population (10,000) that is at risk of being displaced by a potential flood.

Through its mitigation strategy, the jurisdiction determines to reduce the number of residences in the at-risk area, to repair an aging levee, and limit construction in the flood prone area. This is expected to reduce the number of persons at risk from 10,000 to 7,000. It also intends to buyout repetitive damage properties if a flood does occur using federal funding. Its strategy includes modifications to evacuation routes that will reduce the time needed to evacuate 7,000 residents from eight hours to five hours.

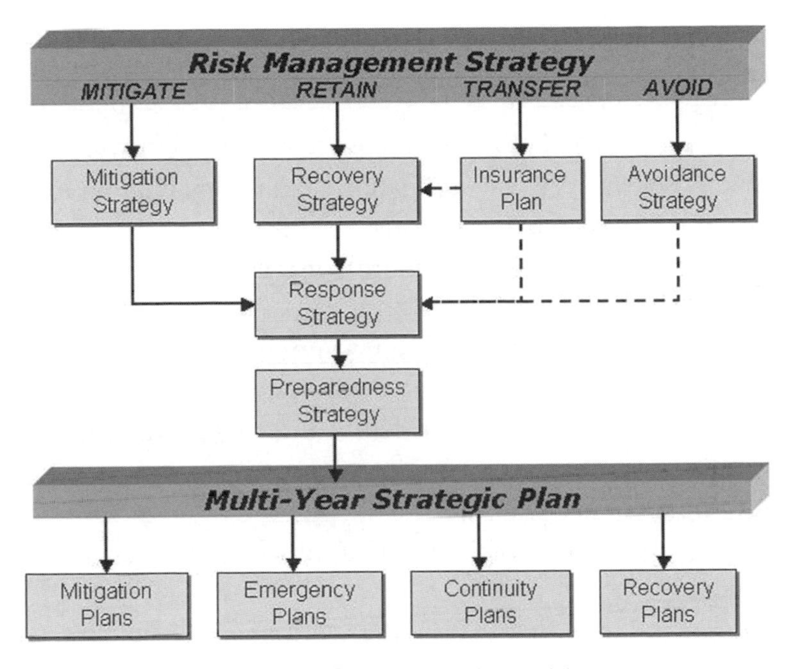

Figure 6.4 Strategic planning model.

In considering recovery strategy, the jurisdiction decides that it needs clearer legislative authority for emergency debris removal, better plans for reentry of residents after evacuation, and estimates that 1,000 citizens will need to be permanently relocated.

As part of its response strategy, the jurisdiction determines that it will need to evacuate and shelter the 7,000 at-risk citizens using a combination of its own shelters and those of an adjoining jurisdiction.

In its preparedness strategy, the jurisdiction determines that it has the capacity to shelter 4,000 citizens and will need to transport 3,000 citizens to the adjacent jurisdiction. It will need to coordinate a sheltering plan with the adjacent jurisdiction. It estimates that it will need interim housing for 3,000 and long-term housing for 1,000. Training will be needed for 1,000 shelter workers and a public awareness campaign will be needed for the new evacuation measures.

In its strategic plan, the jurisdiction identifies the following work elements:

- Propose new legislation for emergency debris removal authority
- Develop elements of the mitigation plan for evacuation routes and the buyout of destroyed property
- Develop a shelter plan and mutual aid agreement with the adjacent jurisdiction
- Develop an appendix to the emergency response plan's shelter annex
- Develop an evacuation plan identifying routes and transportation assets
- Develop a reentry plan
- Develop a public awareness campaign for new evacuation control measures
- Conduct training for 1,000 shelter workers in conjunction with the Red Cross

The intent of the process is to develop response concepts that are based on community vulnerabilities to identified hazards and that are consistent with community values. The process considers emergency management in its broadest perspective: a mechanism by which the community manages its risk. To be effective, this mechanism must consider all the dynamics that are taking place during a disaster and not be limited to traditional emergency response planning. It is the interrelationship among the four concepts of mitigation, preparedness, response, and recovery that makes the CEM model such a powerful strategic tool.

Chapter 7

PLANNING CONCEPTS

... I have always found that plans are useless, but planning is indispensable.
—Dwight D. Eisenhower

The previous chapters have placed considerable emphasis on strategy development as the precursor for developing emergency plans and on the development of a governance structure and program to support the planning process. The emergency management program in essence creates the context in which effective planning can take place. However, strategy must be translated into practical applications that guide tactical and operational response. The ultimate expression of strategy is the emergency operations plan.

The emergency operations plan is not an end in itself. In fact, a plan-centric emphasis has been the downfall of many emergency planners. Without the operational procedures and the logistical and financial structures to support them, emergency operations plans are ineffective. Planning must be inclusive and consider all the elements needed for successful response. This means going beyond the traditional view to accept that emergency operations plans must encompass not only life-safety response but continuity and recovery as well. This chapter looks at some of the concepts and methodologies that help translate strategy into effective emergency response plans.

PLANS VERSUS PLANNING

As noted in the quotation above from Dwight Eisenhower, a fundamental concept of emergency management is that there is a considerable difference between a plan and the act of planning. To many, a plan is a physical document developed to meet a requirement or need, either as the result of legislation or by public demand. It is a tangible object that is often the result of intense effort and is frequently used as a measure of success for a program or as evidence of compliance with laws or regulations.

However, a plan is never more than a snapshot of an organization's intent at a specific point in time. Many plans are out of date almost as soon as they are published. Further, if the written plan is the product of a consultant or single office rather than of a multi-disciplinary task force, the result may be what Erik Auf der Heide in *Disaster Response: Principles of Preparation and Coordination* refers to as the "Paper Plan Syndrome."

The Paper Plan Syndrome refers to the use of a written plan to create the illusion of preparedness. The fact that the organization has created a plan sufficient to meet requirements is considered evidence that the organization has the capability to deal with crisis. This is a frequent problem where plans are mandated for accreditation or by law and organizations produce plans for this purpose with limited concern for valid planning assumptions and follow-on training and exercises. The results of the Paper Plan Syndrome can be devastating, as illustrated in the case study of the Alyeska Oil Spill Contingency Plan in place at the time of the Exxon Valdez disaster.

The Alyeska Oil Spill Contingency Plan was based on a scenario that almost perfectly matched the conditions of the spill. However, the company had written the scenario under protest, did not believe the event could actually occur, and had made no effort to stockpile the resources that would have been needed to implement the plan. However, the existence of the plan created the illusion that the comany was prepared to respond to an oil spill of 200,000 barrels of oil, an

CASE STUDY: ALYESKA OIL SPILL CONTINGENCY PLAN—PAPER PLANS LEAD TO FAILURE

"We have adequate knowledge for dealing with oil spills and improvements in techniques and equipment are continuing to become available through worldwide research. The best equipment, materials, and expertise which will be made available as part of the oil spill contingency plan will make operations at Port Valdez and Prince William Sound the safest in the world..."

—L.R. Beyon, British Petroleum speaking for Alyeska in 1971

The State of Alaska requires oil facilities operating in the state to have an oil spill contingency plan that must be reviewed and approved every three years by the Alaska Department of Environmental Conservation (ADEC). In March 1982, ADEC reviewed a plan that had been developed in 1980 by the Alyeska Pipeline Service Company for potential spills in Prince William Sound. ADEC gave the plan a "conditional approval" because it felt that Alyeska's assumption that the "maximum credible spill" of 74,000 barrels was unrealistic. Many tankers transiting Prince William Sound carried over a million barrels of oil and ADEC believed that a 74,000-barrel spill estimate was too low. Alyeska initially questioned ADEC's jurisdiction over its operations but by June revised its plan and claimed it could recover 100,000 barrels of spilled oil in 48 hours.

In 1986, ADEC required recertification of the contingency plan and requested that the plan include provisions for a 200,000-barrel spill. Alyeska submitted a plan with two scenarios: one for a 4,000-barrel spill that Alyeska believed was the most likely scenario and another for 200,000 barrels. The 4,000-barrel scenario assumed almost optimal conditions, with seas at less than five feet, two miles of visibility, and weather conditions being sunny or overcast with light rain and winds at eight knots. All organizations involved in response were to be notified immediately, booms would be in place within three hours, and an Arizona-based dispersant contractor would be on scene and working within nine to 17 hours. Cleanup was projected to last two months and all but 100 barrels would be recovered or dispersed.

Alyeska strongly objected to the inclusion of the second required scenario and stated in the plan that it believed such a scenario was highly unlikely, citing in part the fact that its ships were American registry and piloted by licensed masters or pilots. Nevertheless, it developed a plan for the 200,000-barrel scenario to meet the requirement for certification. The plan assumed weather conditions identical or better than those in the 4,000-barrel scenario (winds were at five knots), that booms would be in place in five hours around the vessel and "immediiately" around the shore, and that the only 10,000 barrels of oil would be unrecoverable.

(continued on next page)

(continued from previous page)

On March 23, 1989 at 12:04 A.M., the Exxon Valdez, an oil tanker with a cargo of 1.2 million barrels of oil, ran aground on Bligh Reef in Prince William Sound, ripping open eight of its 11 cargo tanks and three saltwater ballast tanks. The damaged tanker eventually released 260,000 barrels of oil into the sound. Weather conditions at the time of the accident were very similar to those anticipated in the scenario: a slight drizzle of rain mixed with snow, north winds at 10 knots and visibility 10 miles. Overall, the accident was an extremely close match to the scenario developed by Alyeska's planners.

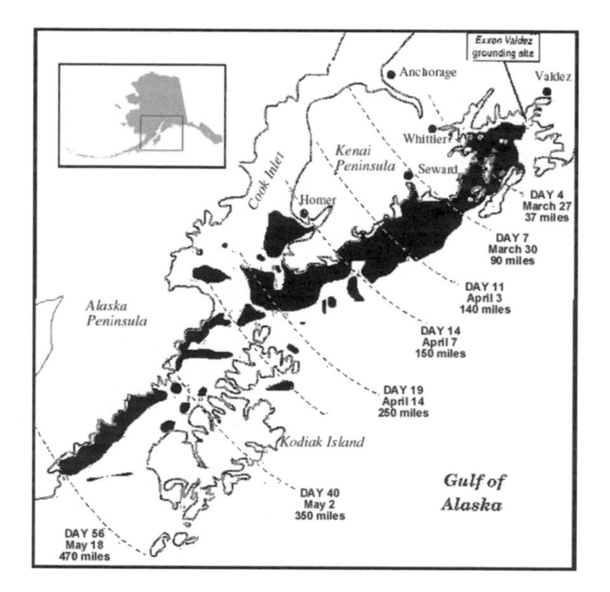

Case Figure 7.1 Map of the Exxon Valdez Spill. Courtesy of the Exxon Valdez Oil Spill Trustee Council.

Alyeska's response, however, was characterized by confusion and delays and soon demonstrated that the contingency plan was little more than a paper document. A contingency barge that state officials believed was loaded with containment equipment in accordance with the contingency plan was empty. Employees and contract laborers had to locate cleanup equipment in warehouses and load them aboard vessels before deploying. Rarely used equip-

(continued on next page)

(continued from previous page)
ment, such as deep-water skimmers and booms, were buried under more frequently used heavy equipment. Some equipment could not be located—heavy ship's fenders vital to offloading oil to another vessel were buried under snow. The barge finally left the terminal at 11:00 A.M. and the first booms were not deployed until sometime between noon and 5:00 P.M. Ironically, dispersants were ineffective because of the calm conditions at the site, but since the amount on hand was too little to deal with such a large spill, their effectiveness would have been limited in any event. Likewise, there were insufficient skimmers and booms on hand for the size of the spill. The resources necessary to implement the 200,000-barrel spill contingency plan that might have limited the scope of the disaster had never been put in place.

Case Figure 7.2 Exxon Valdez and large oil slick in Prince William Sound. Courtesy of the Exxon Valdez Oil Spill Trustee Council. EVOSTC photo by Erich Gundlach.

Eventually, the spill would have an impact on 1,300 miles of shoreline and 40,000 people. Cleanup operations would involve 10,000 workers, 1,000 boats, and 100 aircraft and cost Exxon an estimated $2.1 billion. Impact on wildlife was estimated at 250,000 seabirds, 2,800 sea otters, 300 harbor seals, 250 bald eagles, up to 22 killer whales, and billions of salmon and herring eggs killed.

(continued on next page)

(continued from previous page)

Damage to the region's tourism industry was estimated at $2.8 billion. Exxon eventually paid over $1.1 billion to settle criminal charges and civil suits with the United States and the State of Alaska. As of 2002, a number of private civil suits (estimated at $5.3 billion) were still unresolved.

An EPA report prepared for the President following the spill, commenting on plans at all levels, sums up many of the issues with Alyeska's Oil Spill Contingency Plan for Prince William Sound:

> *Exxon was not prepared for a spill of this magnitude--nor were Alyeska, the State of Alaska, or the federal government. It is clear that the planning for and response to the Exxon Valdez incident was unequal to the task. Contingency planning in the future needs to incorporate realistic worst-case scenarios and to include adequate equipment and personnel to handle major spills. Adequate training in the techniques and limitations of oil spill removal is critical to the success of contingency planning. Organizational responsibilities must be clear, and personnel must be knowledgeable about their roles. Realistic exercises that fully test the response system must be undertaken regularly.*
>
> *The Exxon Valdez Oil Spill: A Report to the President (Executive Summary), Environmental Protection Agency, May 1989.*

illusion that was accepted by the State of Alaska's Department of Environmental Conservation (ADEC) who approved the plan.

Further complicating the issue of planning is the large variety of plans with similar sounding names and purposes: response plans, emergency plans, mitigation plans, business continuity plans, disaster plans, crisis management plans, etc. Although many of these plans are similar and are intended to coordinate an organization's response to crisis, they each have a different emphasis and, in many cases, a different promulgating group within a jurisdiction. Because the names of the plans are not precise and lack a common definition, the plans may overlap authorities or give organizational groups multiple assignments. The result can be confusion and a tacit agreement to ignore the plans and do what seems right.

This tendency to ignore plans is far from unusual. Since many plans are intended to be read by a program evaluator rather than by a user, plans tend to be bulky and wordy and are frequently organized in a way that makes information hard to find. This apparent disorganization is partially the result of the heavy influence of military planning on emergency management. Actually, these plans are incredibly organized and the collection of annexes, appendices, tabs, etc. represent very detailed planning. However, this type of planning format was designed for strategic level plans and was never intended for use by troops in direct contact with the enemy.

In a time of crisis, one is lucky if a responder even thinks to make use of the checklists contained in many plans. Following the 1980 eruption of Mount St. Helens, Washington State University studied 26 communities affected by the event and found that the majority had not used an emergency plan, with many respondents stating that the plan they did have had not been applicable to their needs. A study of 29 mass casualty disasters conducted by Quarentelli in 1983 also found that written plans were implemented in only a small percentage of the cases studied.

On September 11, District of Columbia police lacked an evacuation plan. The need for such a plan had been identified in July 2001, but the planning process was just underway when the attack on the Pentagon took place. Instead, the police began cobbling together a response plan based on existing procedures and plans. A traffic plan devised for the millennium was used to clear downtown intersections in about three hours. Recall procedures used during presidential inaugurations were used to recall off-duty officers. Crowd control was handled using procedures established for the civil disturbances that are common in Washington. The result was not elegant and not the optimum solution, but District police were able to respond with some degree of effectiveness in the absence of a plan.

There is evidence to suggest that plans may, to a certain extent, be counter-productive. In *Managing the Unexpected*, researchers Karl Weick and Kathleen Sutcliffe found that the techniques used by many planners can foster a mindset that leaves planners vulnerable to the unexpected. Traditional planning uses stable and predictable contexts, giving the impression that a disaster will unfold in a predetermined manner. Plans can therefore influence the perception of the disaster by creating expectations that may not be based in reality. If an event does not fit the concepts of the plan, it may not be recognized as the first stage of a disaster. This is consistent with the natural human tendency to normalize occurrences and treat them as routine. People see what the want to see or what they have been conditioned to see.

Plans can also have a limiting effect on response. Planners make assumptions about what is likely to occur and produce detailed contingency plans to meet those assumptions. There is a tendency to start twisting perceptions of the disaster to fit these contingencies; that is, to see what the plan says should be there, rather than making an unbiased assessment. This distortion also applies to an assessment of capability—to a certain extent plans condition responders to think in certain patterns regarding how capabilities are used. Improvisation is not encouraged.

Another concern is the expectation that following the established patterns in the plan will automatically lead to success. This may be true if the planners have foreseen every circumstance, but as the Exxon Valdez case study demonstrates, even where the plan is consistent with the actual event, there are no guarantees of success. Disasters are characterized by the unexpected, creating new tasks and bringing new players to the table. Emergency plans should therefore place a high premium on improvisation and creative problem solving.

This is not to say that plans should not be written, but rather that the traditional plan-centric approach is not effective. Plans are not an end product; they are merely one component of the emergency management program. Written plans do serve a purpose within the program. They document the measures that a community has put in place to deal with risk and can provide continuity in organizations where turnover is high. They represent a form of closure for the planning team, providing tangible proof that something is actually being accomplished. They can be valuable training tools and serve as standards to guide the development of other components of the program.

However, the written plan is never as important as the process that is used to develop it. A plan that does not represent an organizational consensus, is not kept reasonably up to date, and is not used as the basis for training and exercise is what Lee Clarke in *Mission Improbable: Using Fantasy Documents to Tame Disaster* terms a "Fantasy Document"—a document that has no connection to the organization's actual capabilities and is little more than a description of what the organization envisions happening after a disaster. It represents a desired outcome but does not necessarily guarantee that this outcome will be achieved.

Once one abandons the plan-centric approach, it is now possible to construct a planning process that is more effective and responsive to the needs of the community. Instead of providing the type of detailed command and control instructions that have been the hallmark of traditional plans, one can now focus on coordination. Effective plans facilitate creative problem solving. They establish an operating structure that can be expanded to include new actors, provide for supporting structures such as effective communications flow, and eliminate potential barriers to improvisation. They are a collection of tools and options that can be combined in new ways to meet the unique needs of each disaster.

THE PLANNING CONTINUUM

NFPA 1600 identifies five basic plans that an organization is expected to have. The first of these, the strategic plan, was addressed in previous chapters. The four other plans are:

- An emergency operations/response plan that assigns functional responsibilities for actions to be taken in an emergency.
- A mitigation plan that establishes short and long-range actions to eliminate or reduce the impact of hazards.
- A Recovery Plan that develops priorities and strategies for restoring the organization's services, programs, facilities, and infrastructure.
- A Continuity Plan that identifies critical functions that must be maintained while the organization recovers from a crisis.

This requirement seems straightforward until one begins to realize that these plans are frequently prepared by different organizational elements that do not always deal with each other on a day-to-day basis. For example, response plans are normally the purview of the local Office of Emergency Services, while mitigation is the province of the planning department. In a private company, business continuity planning may be a function of an information technology department while risk management is part of the financial department. This separation of responsibility is further complicated by the overlap of responsibilities—where, for instance, does continuity end and recovery begin? Does the emergency operations center have a function in recovery? Is mitigation a priority during response?

These problems are generated because organizations tend to be plan-centric. The focus is on the end product—the paper plan—and not on the overall process. The planning process is stove-piped by involving only those stakeholders traditionally thought to be responsible for the specific planning issue. There is little emphasis given on creating a holistic approach that makes use of all the organization's capabilities because there is little understanding of the distributive nature of emergency management discussed in previous chapters.

The first step to approaching planning holistically is to accept that planning functions are not discrete units that can be placed in unique time frames. The Comprehensive Emergency Management Model discussed in previous chapters is intended to demonstrate the interrelated components of emergency management and does not reflect actual progression over time. Plans, therefore, are not implemented sequentially, but simultaneously. This means that they are interlinked in a continuum that begins before the event and extends into long-term restoration. Within this continuum, components of various plans are implemented over time as necessary.

This simultaneous implementation must be coordinated to ensure that one plan's activities do not have an adverse impact on another's. For example, the need to demolish unsafe structures as soon as possible may prevent mitigation teams from assessing the performance of pre-disaster mitigation measures. Similarly, the need to jump-start the economic recovery of businesses may affect priorities for power restoration. Consequently, emergency planners must consider the need for managing the simultaneous implementation of multiple plans.

A useful concept in implementing holistic planning is the understanding of the progression of plan development. While most plans are actually developed simultaneously, viewing plans in a conceptual

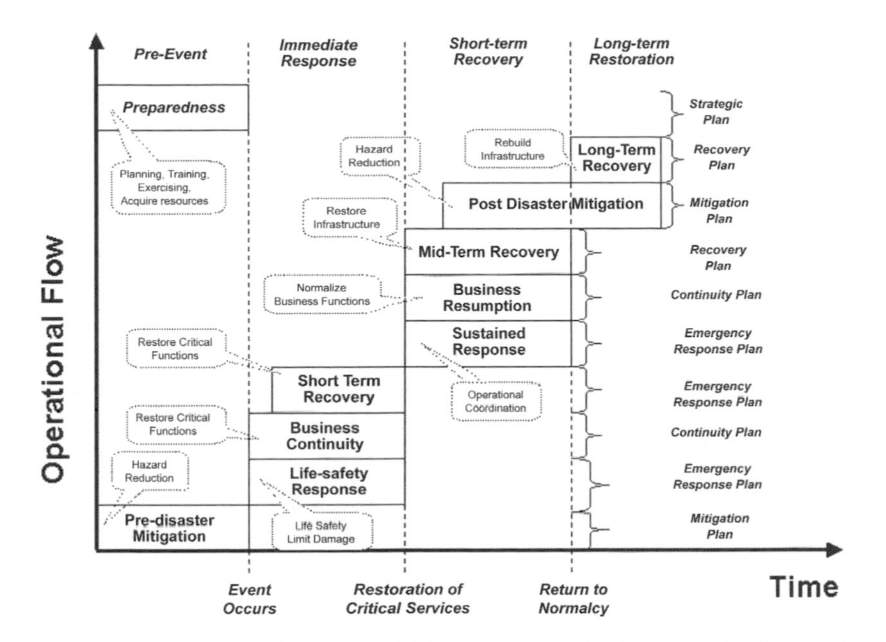

Figure 7.1 Planning continuum—multiple plans may be implemented simultaneously, creating confusion over priorities and competition for scarce resources.

sequence helps to understand how they are interlinked. Plans follow a natural sequence based on the needs they address, much as Maslow's "Hierarchy of Human Needs," a basic management concept, addresses progressive human needs. Before a plan at a certain level can be truly effective, lower level plans must be in place to support it. For example, mitigation plans can affect the anticipated level of response operations required in the emergency operations plan. Likewise, failing to plan for immediate life-safety response makes continuity and recovery planning moot. Figure 7.3 shows a model of this "plan hierarchy."

As has been previously discussed, emergency planning is based on an assessment of the hazards facing the community. The community's tolerance for risk determines the mix of strategies that will be used to

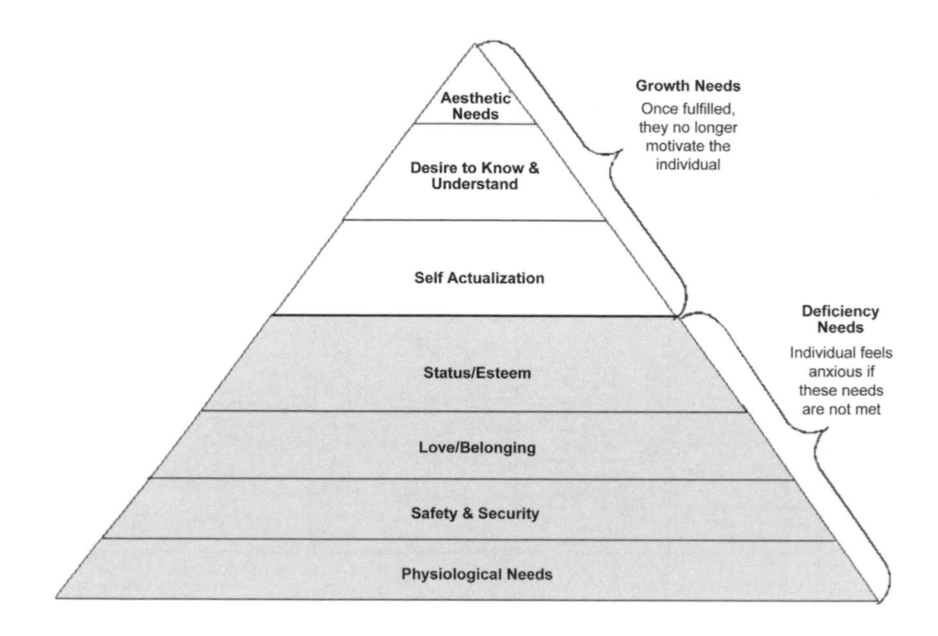

Figure 7.2 Maslow's Hierarchy of Needs—based on Abraham Maslow, *A Theory of Human Motivation* (1943) and *Motivation and Personality* (1970). Although it has its detractors, Maslow's theories were considered ground breaking in understanding human motivation and were used extensively in managerial training.

manage risk and these strategies ultimately produce a strategic plan that calls for the development of various other plans and supporting documents.

Although response-oriented emergency managers often overlook it, mitigation is of critical importance to the community. Mitigation is taking sustained actions to reduce or eliminate long-term risk to people and property from hazards and their effects. If mitigation is successful, the need for response and recovery actions are either eliminated or reduced and the community is able to return to normalcy quicker with limited operational disruption and at a lower cost.

Figure 7.3 Plan hierarchy—while not a direct analogy to Maslow's Hierarchy, plans can also be ranked in such a way that suggests that lower level plans must be addressed before higher-level plans can be effective.

Not all risks can be prevented, however, and there is always the potential for an unforeseen or low frequency event to occur. Consequently, the community needs to develop a capacity for immediate action in the event of an emergency. The first priority of any response is always life-safety. For this reason, most emergency operations plans deal with issues such as rescue, immediate medical attention, sheltering, and so forth. However, while personal survival is a priority, organizational survival must also be a priority. Where emergency operations plans provide for the survival of individuals, continuity plans provide for the survival of government as a business entity. For this reason, emergency operations plans and continuity plans are generally implemented simultaneously. Both are intended to bridge the gap between the event's occurrence and the transition to long-term recovery operations.

Immediate response plans, whether aimed at life safety or continuity, are intended to maintain the critical functions of the community. They are designed to eliminate or reduce immediate threats to life-safety and to enable the continuity of critical operations that allow the local government to perform essential functions. Essentially, they provide a quick fix to immediate problems. Recovery plans address long-range issues that have as their goal a return to normalcy for the organization. This period can be extremely lengthy and may involve strategic decisions regarding the future of the organization. Recovery represents an opportunity to make substantial changes in how the organization achieves its goals and developing a mechanism to identify these opportunities and guide decision-making is a critical part of recovery planning.

Understanding the relationship among the elements of the plan hierarchy is a critical part of emergency planning. If the previous elements have not been adequately addressed, this will have a significant effect on higher levels:

- If the hazard analysis is flawed, the community may not adequately prepare for potential problems.
- If the strategic plan has not been developed, plans may not be synchronous and critical planning issues may not be addressed.
- A poor or non-existent mitigation plan means that the impact of the disaster will be greater on the community, increasing the demands on personnel and resources coordinated under the emergency operations.
- Inadequate emergency plans result in chaos, limiting the community's ability to manage response and drawing resources from continuity operations.
- Failing to plan for continuity means that recovery operations may be delayed or may have to be curtailed because the community is unable to raise sufficient capital for essential functions and rebuilding.
- Recovery plans directly affect the speed and direction of reconstruction and may determine the survivability of the community.

CENTRALIZED PLANNING, DECENTRALIZED EXECUTION

So far this discussion has treated the five basic plans identified in NFPA 1600 as if they were single plans. In some cases, this may be true but normally plans are multi-leveled. That is, sub-elements of a community may have separate plans that are integrated within an over-arching jurisdictional plan. These individual plans may be annexes or components of the jurisdictional plan or may be specific to a department or agency. Supporting plans can be incorporated into the main jurisdictional plan by reference or can be listed as annexes or appendices to the plan. Further, NFPA 1600 requires the development of operational procedures that may be incorporated into standard operating procedures or protocols that also form components of the plan.

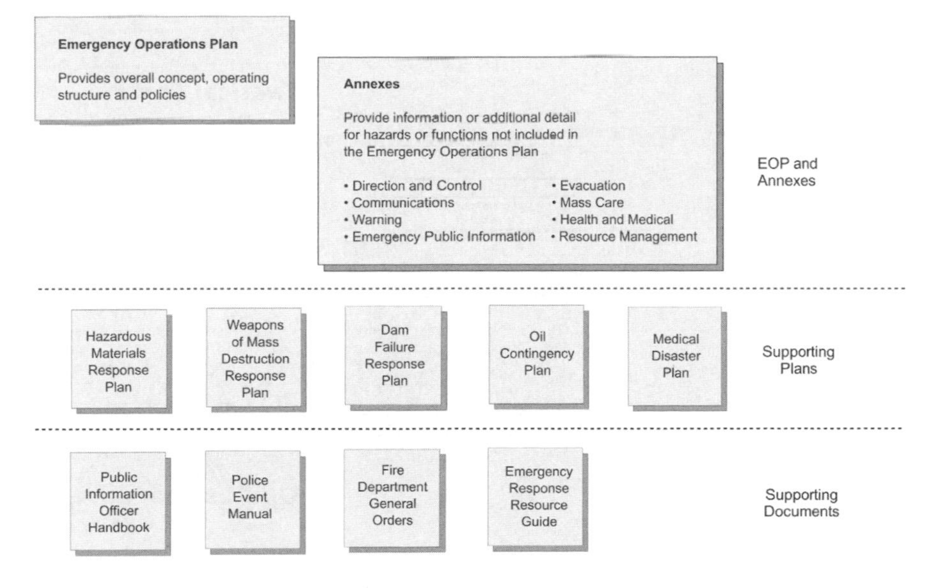

Figure 7.4 Emergency plans consist of more than a single emergency operations plan.

The level at which a plan is set has a direct bearing on how that plan is developed and written. Generally, the more specific a plan is, the lower the level at which it is intended to be used. Community-wide plans tend to be tactical in nature—they discuss overall concepts, provide planning guidance, and establish structures for assessment and decision-making. Supporting plans are primarily operational—they are narrow in scope and tell the users specific information about how to implement the plan. Plans at the full operational level may be published as field operating guides and include checklists that distill the plans down to essential tasks. Figure 7.5 shows an example of how a jurisdictional plan may be supported by plans at different levels. In this example, a centrally developed jurisdictional shelter plan is implemented through a series of department plans, procedural documents and checklists.

An understanding of this concept of plan levels can prevent the inclusion of extraneous material in tactical plans that provide too great a level of detail for the user. Likewise, it can help focus operational level procedures by excluding unnecessary conceptual information while ensuring the necessary level of detail for the user.

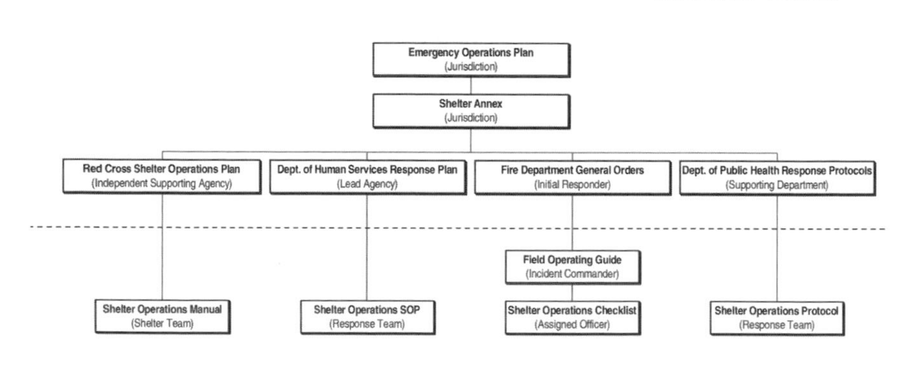

Figure 7.5 Synchronicity in planning—a centrally developed tactical plan is used to coordinate the operational plans of several agencies to achieve an agreed upon result.

Plan implementation is decentralized. Agencies may have autonomy in implementing their individual plans owing to the importance of immediate response or the possibility of communications failures. For example, fire officers may institute an incident command structure to respond to an observed crisis. Continuity plans are implemented without central authority. Other organizations may have specific triggers to begin operations. Many community emergency response teams (CERTs) are organized to begin operations as soon as a specific event such as an earthquake occurs.

It is also not unusual to see a transition in leadership over time as the need for specialized expertise changes. For example, agencies responding to an explosion may initially be managed by the fire department because of their life safety responsibilities. Command can rapidly transition to the police to coordinate the criminal investigation and then to public works for demolition and site clean up. The tasks required of each agency are still being performed simultaneously, but overall coordination is handled by the agency best suited to do so at that particular point in the operation. The unifying factor in the response is the centrally developed tactical plan, which would include an agreed upon incident management system.

Various responses to the same incident may also be treated as separate operations by the agencies involved. Referring back to Figure 7.5, the fire department responding to a large fire rescues victims and moves them to a staging area where they receive immediate medical treatment. Responsibility for the victims then passes to a Red Cross team that does an initial assessment of victim needs and provides emergency sheltering. As part of the sheltering operation, city services assess unmet needs and the victims are enrolled in social programs as required. From the fire department's perspective, the incident is the fire and their work in relation to sheltering is done once the victims have been staged and received medical attention. For the Red Cross, the incident is the shelter operation, with the fire scene assessment being an initial phase of the overall operation. To the city social services agency, the provision of relief is a relatively routine operation, with only the assessment and intake at the shelter requiring an unusual field deployment. The three separate autonomous responses constitute a single coordinated community response to the fire.

The development and implementation of multi-level plans with decentralized execution depends on coordinated, centralized planning. Centralized planning does not address all the specifics needed for lower-level plans nor does it mandate the level of detail required in those plans. These details are left to the subordinate groups responsible for developing operational level plans. Centralized planning does not necessarily mean central approval either. Rather, it presupposes a unity of concept that ensures that all plans are written with a common purpose in mind, even as supporting plans also address the responsibilities of each planning group. Centralized planning articulates the overall end result that the organization hopes to achieve through its planning.

This unity of vision, or unity of intent, highlights the importance of the strategic planning model discussed in the previous chapter. To adequately manage centralized planning, it is absolutely critical that the community have clearly articulated strategies to guide the planning process. Issues related to conceptual approaches, policy, governance, and capabilities should have been raised and addressed as part of strategy development, allowing planners to focus on operational implementation of the strategies.

PROBLEM SOLVING MODEL

In a certain sense, plans represent anticipated solutions to potential problems. A weakness in the development of many plans is the failure to use a structured problem-solving methodology as a framework for plan development. Figure 7.6 shows two problem-solving models, one from the National Management Association and the other from the U.S. Army. Note that the two models are almost identical, as business and the military have traditionally borrowed concepts from each other. The successful use of a similar model in both spheres suggests that problem solving should follow a logical and orderly progression. This progression does not seem to be intuitive, however, as many organizations spend considerable time spinning their wheels in "planning meetings" without producing an outcome.

The most underused step in the problem-solving model is the most basic one: define the problem. Without a clear definition of what the

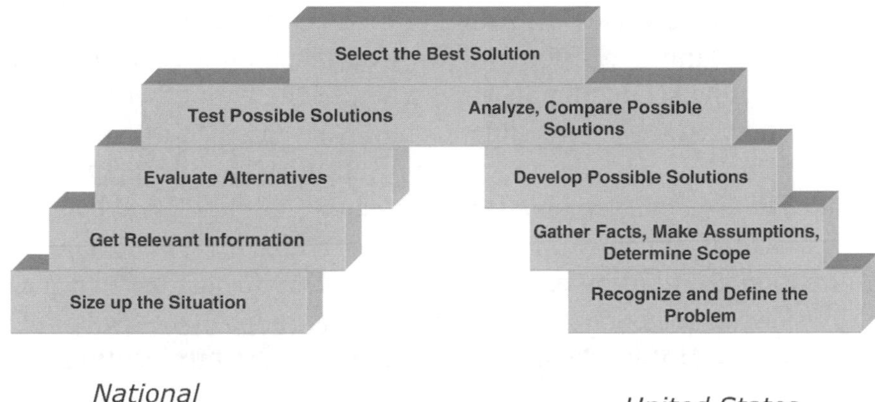

Figure 7.6 Problem solving models—models used by the private sector and the military show remarkable congruence. While the approaches vary slightly, the methodology and the outcome are the same.

problem is, it is easy to be distracted by tangential issues. For example, in response to a series of residential hotel fires, a major jurisdiction formed a task force to provide legislators with recommendations on how to better mitigate fire hazards and how to stimulate reconstruction of damaged hotels. However, owing to poorly crafted enabling legislation that expanded the scope of the task force to address tenant welfare, such issues as visitation policies, elevator repairs, and family housing sidetracked the task force. Eventually, many members of the task force who were committed to the original mission felt they were being misused and began to miss meetings. Worse, since the task force also involved itself in response issues, an existing task force on sheltering chaired by the local emergency manager saw a significant drop in meeting attendance as members became confused on the roles of the two task forces.

Once one understands the planning continuum and the concepts of centralized planning, problem definition takes on a critical role in determining the scope of a plan and where it fits with other plans. By defining this scope, it is possible to state what the plan is expected to accomplish and, perhaps more importantly, what the plan is not intended to do. For example, if the plan addresses high-level concepts (i.e. is tactical in nature), it should not contain the type of detailed information required by a field responder. This means that the planning team can focus on developing concepts rather than spend valuable time assembling detailed procedures not directly useful to the plan.

In the same vein, if all plans are written at the tactical level, it is unlikely that anything will actually be accomplished by the organization. Again, if the plan is intended to be an operational plan, a clear definition of scope is essential to make sure that concepts are kept to the minimum needed for plan implementation and the plan is heavily weighted toward the type of detailed information needed to accomplish its intent.

PLANNING ASSUMPTIONS

Also frequently overlooked in planning are the collection of relevant data and the making of assumptions. Relevant information does not have to be the result of exhaustive research, but it should at least cover facts that bear directly on the plan. For example, if one is developing an evacuation plan for a facility, it would be important to know the number of personnel who work at the facility, the number of visitors, and whether these numbers vary by time of day or by time of year. Information of this sort can be gleaned from a number of sources, such as historical data or academic studies. One can also make assumptions about the availability of response resources or about the time it will take to receive mutual aid assets.

Planning assumptions are essential to developing any plan. No matter how much information is gathered, there is no predicting exactly which crisis will occur or the exact nature of the crisis. Consequently, one must make assumptions on what is possible or likely to occur in order to define the scope of the plan. One of the lessons suggested by the Alyeska case study is that changing assumptions is not necessarily just a question of degree. This is reinforced by the research by Quarantelli and others that note qualitative differences between emergencies, disasters and catastrophes. Each change in assumptions has the potential to change planning parameters substantially and not just by degree. Therefore, assumptions made in planning must be clearly identified and validated.

Planning assumptions should be reality-based. That is, they should, as much as possible, reflect actual disaster experience and a realistic appraisal of the conditions under which the response will take place. For example, the Association of Bay Area Governments in California has produced forecasts of the impacts of a major earthquake on regional transportation systems and housing stocks in the San Francisco Bay Area. These forecasts allow for realistic assumptions about the survivability of transportation infrastructure and about the potential for displaced populations. Similarly, models such as FEMA's Hazards US (HAZUS) modeling program can offer useful planning data, provided one is familiar with the limitations of such models.

Not all assumptions based on past performance are relevant, however. The City of New Orleans based its communications plans on an assumption that the telephone system would continue to function in a disaster, an assumption that in retrospect seems completely out of touch with the realities of Hurricane Katrina. However, the system had worked in all previous disasters and the planners considered the assumption valid. Previous actual experience had lulled them into a position where they ignored the experience of other Gulf Coast states and the projections of a number of researchers and government agencies.

Planners need to guard against unspoken assumptions as well. These are assumptions that are considered self-evident or are not even realized as being assumptions. There is an unspoken assumption in most plans, for example, that the jurisdiction will survive with most of its ability to coordinate response intact. This is not necessarily an unreasonable assumption and could produce a solid baseline plan with alternate scenarios being addressed through exercises. However, if the community has not given thought to continuity planning, this assumption may not be valid. Returning again to the Alyeska case study, it was clear that assumptions were made about the equipment that was stockpiled at Valdez and its level of readiness. Regardless of how generally accepted they may be, underlying planning assumptions should be identified as part of the planning process.

SCENARIO-BASED PLANNING

The development of assumptions is part of a planning technique known as "scenario-based planning" that forms the basis of many plans. Originally developed in the 1960s to analyze military strategies associated with thermonuclear war and epitomized by Herman Kahn's book *Thinking the Unthinkable*, scenario-based planning was quickly adopted by the business community. Its success as a business-planning tool was highlighted by the work of Pierre Wack, whose scenario planning system positioned Shell Oil to weather the oil crisis in 1973. It continues to be a mainstay in planning in both the public and private sectors.

Scenario-based planning requires the identification of predetermined elements that are virtually certain to occur, then coupling them with those elements that could possibly occur to craft potential scenarios of what might occur. The impact of each scenario and the associated response are then analyzed to determine potential strategies. For example, many jurisdictions are subject to floods and there is no question that a flood will occur at some point in the jurisdiction's future. The task of the planners is to study available data and determine the various types of floods that may be faced by the jurisdiction, which type is most likely to occur, and the worst-case flood. These scenarios can then be used to make decisions about response strategies and resource shortages.

Scenarios can be constructed and used in a number of ways:

- Worst-case scenario—during the hazard analysis, planners identify the event or events having the greatest impact on the community. The resources needed to respond to this event or events are then determined and the emergency management program is geared toward reducing these shortfalls.
- Most-likely scenario—the worst-case scenario for many communities is usually a catastrophic event. As Hurricane Katrina has demonstrated, this means that the ability of local government to

centrally manage the event may be severely degraded or completely disrupted. Consequently, planners may opt to use a scenario or scenarios that are most likely to occur or that represent the maximum credible event with which the local government can cope.

- Impact-based scenario—Instead of focusing on a specific event, planners study the impacts of various hazards to construct a scenario based on the maximum impacts the community may face. For example, power outages are to be expected during disaster. The length of the outage will vary based on the scenario. A planner developing an impact-based scenario would select the maximum expected outage for use in the scenario. In essence, an impact-based scenario is a collection of assumptions based on the outcome of the hazard analysis to produce a generic worst-case scenario based on impacts rather than hazards.

Scenarios are particularly useful in helping to determine planning assumptions. Making assumptions is a balancing act. Conventional wisdom suggests that if one can deal with the worst-case scenario, one should be able to deal with lesser events. However, selecting the worst-case scenario may change the parameters of the plan by making the scope of the plan too large for a realistic response. This may actually deter planning, as the community may be unwilling to commit the necessary resources to deal with such a large-scale event.

This issue provides a dilemma for the emergency manager. As was noted in Chapter 5, catastrophic events and major disasters have such a low incidence that it is difficult to gauge probability of occurrence. This suggests that, if an event has occurred historically, no matter how remotely, it should be at least considered in planning. In other words, the careful analysis described in Chapter 5 should not be considered an exact science but must be supplemented by good judgment. The probability of a major terrorist attack cannot be accurately calculated because of the low incidence, but the risk is real enough that emergency planners have to consider it.

However, some incidents, such as some of the geologic upheavals of the past, are considered to be of such low incidence that they fall outside a community's tolerance for risk. This means that the jurisdiction considers the event so unlikely that it is not prepared to expend resources to prepare for it. In essence, the jurisdiction turns a blind eye to the possibility of the event occurring. This tolerance for risk must be factored in as one selects suitable planning scenarios. There are occasions where a worst-case event is appropriate and necessary to demonstrate the need for more resources but, in many cases, the scenario selected will be the most likely one or the maximum-credible event to which the jurisdiction can or is willing to respond.

In selecting the most likely scenario, however, there is the potential to ignore less probable events with potentially high impact on the community, providing a false sense of security in the community's ability to handle crisis. This is clearly illustrated in the Alyeska case study. The planners selected a best-case scenario in regard to the weather and sea conditions that was, in retrospect, completely unrealistic. A worst-case scenario, with a vessel breaking up in the midst of a major storm, would have substantially changed the plan, as no containment of the spill would have been possible. However, the scenario that was selected contained conditions that were ideal from the planners' perspective but were rare in Prince Edward Sound. Gathering historic information on weather and sea conditions might have allowed planners to make assumptions as to what conditions were most likely to occur and this would have been a more reasonable basis for a plan.

Scenarios are particularly useful in developing contingency plans for specific occurrences. Contingency plans are usually developed for events that can be anticipated and have limited variables. They tend to be specific in nature and primarily operationally oriented. An example of this type of plan is an evacuation plan associated with a flood inundation zone below a dam. A community with multiple dam sites could prepare a contingency plan for each site providing pre-designated routes and operating facilities (e.g. staging areas, shelter sites, etc.). Units responding to an inundation would operate under the overall jurisdictional plan and department plans and guidelines and would use the contingency plan to begin operations quickly. In essence, the contingency plan serves as an initial Incident Action Plan, saving time in a critical situation.

Recently, the Department of Homeland Security has begun a push toward scenario-based planning through the introduction of a series of

National Planning Scenarios (Figure 7.7) aimed at identifying the scope, magnitude and, complexity of major events that could potentially affect the nation. However, the scenario list reflects the overwhelming DHS emphasis on terrorist attacks. Of the 15 scenarios, only two deal with natural hazards: one for a major earthquake and one for a major hurricane. A third scenario is based on influenza pandemic. The remaining 12 scenarios are all terrorist attack scenarios, primarily chemical and biological based.

The problem with such an approach is that in place of a plan-centric system, jurisdictions may evolve to a scenario-centric one. There is a very real possibility that jurisdictions will adopt the scenarios without regard to actual risk thereby using scarce resources to plan and prepare for events that are extremely unlikely to occur at the expense of plan-

Scenario 1: Nuclear Detonation – 10-Kiloton Improvised Nuclear Device

Scenario 2: Biological Attack – Aerosol Anthrax

Scenario 3: Biological Disease Outbreak – Pandemic Influenza

Scenario 4: Biological Attack – Plague

Scenario 5: Chemical Attack – Blister Agent

Scenario 6: Chemical Attack – Toxic Industrial Chemicals

Scenario 7: Chemical Attack – Nerve Agent

Scenario 8: Chemical Attack – Chlorine Tank Explosion

Scenario 9: Natural Disaster – Major Earthquake

Scenario 10: Natural Disaster – Major Hurricane

Scenario 11: Radiological Attack – Radiological Dispersal Devices

Scenario 12: Explosives Attack – Bombing Using Improvised Explosive Devices

Scenario 13: Biological Attack – Food Contamination

Scenario 14: Biological Attack – Foreign Animal Disease (Foot and Mouth Disease)

Scenario 15: Cyber Attack

Figure 7.7 National planning scenarios—note the emphasis on attack scenarios versus natural disasters. This highlights one of the problems with scenario-based planning—selecting scenarios without hazard analysis can skew planning efforts.

ning for those that constitute a very real risk to the community. Worse, jurisdictions may opt to plan for only the national scenarios in order to fulfill grant requirements while ignoring other potential risks.

Scenario-based planning is a powerful tool but it has inherent problems. There are potentially unlimited scenarios for any community. Developing full plans for a variety of potential scenarios is a self-limiting proposition because of the work involved in plan development and maintenance. Further, if an event occurs that does not match a scenario, existing plans may not be suitable. There is a very real danger that the emphasis on specific scenarios may drive the types of behavior identified by Weick and Sutcliffe: creating unrealistic expectations of how disasters will unfold and creating an expectation that following the plan will automatically lead to success. For this reason, planners have moved over the years to multi-hazard functional plans.

FUNCTIONAL PLANNING

The drawbacks of scenario-based planning can, to a certain extent, be offset by the use of functional planning. Functional planning looks at those tasks the community must perform in a time of crisis and attempts to define commonalities. In a sense, this negates the need for separate scenario plans by assuming that functions will remain constant in most crises. It does not matter why sheltering is needed—the process for establishing and running shelters, for providing services, and for transitioning to temporary housing remain essentially the same. Locations, specific needs, and timing—these are variables that are event-specific and will require creativity in the implementation of the plan. However, the approach to sheltering, the resources, and the governance structure all would remain constant despite the nature of the disaster.

Functional planning is an attempt to construct an all-hazards approach to emergency planning and has been the standard methodology used in the United States for many years. It is the basis of *State and Local Guide 101* and many of the FEMA instructional materials. Figure 7.8 shows the eight basic functions listed in SLG 101 that form a logical starting point for functional planning. EMAP recommends a more expanded list of activities (Figure 7.9), some of which can be combined under a single function. For example, sheltering and mass care are usually considered a single function as are resource management and food, water, and commodities distribution.

While the EMAP list is fairly representative of the tasks that must be accomplished in a crisis, the final determination on how the tasks are implemented should be based on the jurisdiction's organizational structure and available resources. In a large jurisdiction with many response resources, each of these tasks may represent a single function. In a smaller one, multiple tasks must be combined under a single function. However, the core of functional planning is 1) the definition of the functions to be performed and 2) the identification of an agency with responsibility for performing the function.

This functional responsibility implies two areas of consideration for the agency. The most obvious is that they have responsibility for accomplishing the function during a crisis. However, if one supports

- Direction and Control
- Communications
- Warning
- Emergency Public Information
- Evacuation
- Mass Care
- Health and Medical
- Resource Management

Figure 7.8 The eight basic functions from *State and Local Guide 101*. These functions form the basis of functional-based emergency response planning.

- Direction/control and coordination;
- Information and planning;
- Detection and monitoring;
- Alert and notification;
- Warning;
- Communications;
- Emergency public information;
- Resource management;
- Evacuation;
- Mass care;
- Sheltering;
- Needs and damage assessment;
- Military support;
- Donated goods;
- Voluntary organizations;
- Law enforcement;
- Fire protection;
- Search and rescue;
- Public health, medical, and mortuary services;
- Agriculture;
- Animal control/management;
- Food, water and commodities distribution;
- Transportation resources;
- Energy and utilities services;
- Public works and engineering services; and
- Hazardous materials.

Figure 7.9 An expanded list of emergency response functions from the Emergency Management Accreditation Program.

the enterprise approach to emergency management, the agency also has responsibility for planning for that function's accomplishment. This suggests that while the agency has responsibility for making sure the function is performed, it does not necessarily need to do so alone. The agency may, in fact, draw upon the resources of the jurisdiction as needed, so long as its planning is centrally coordinated and does not represent a competition for resources.

Consider the issue of evacuation as an example. Within this function are a number of implied tasks: traffic control, route reconnaissance, establishment of fuel and rest stops, activation of reception centers, warning, public information, etc. Each of these tasks is performed by separate agencies and may in itself be a function under the plan. It is clear that the single lead agency cannot handle all these tasks on its own, so as it plans for evacuation, it must identify supporting agencies that can provide these services. The lead agency develops the plan with input from these supporting agencies and at the time of the crisis, coordinates the implementation of the plan. This plan is usually incorporated in the emergency operations plan as an annex and supported by individual agency plans and standard operating procedures.

An excellent example of functional planning is the National Response Plan, in which the federal government has identified the various functions necessary for providing crisis response and assigned each function to a lead agency with other agencies identified for support. The initial phase of a response is a rapid assessment of need by each function and a decision on what resources of each agency will be deployed. Figure 7.10 shows a matrix of the 15 emergency support functions identified in the National Response Plan and the primary and supporting agencies assigned to each function.

The functional approach is not limited to emergency operations plans. As one develops continuity and recovery plans, the same mech-

	Emergency Support Functions														
Agency	#1 - Transportation	#2 - Communications	#3 - Public Works and Engineering	#4 - Firefighting	#5 - Emergency Management	#6 - Mass Care, Housing, and Human Services	#7 - Resource Support	#8 - Public Health and Medical Services	#9 - Urban Search and Rescue	#10 - Oil and Hazardous Materials Response	#11 - Agriculture and Natural Resources	#12 - Energy	#13 - Public Safety and Security	#14 - Long-term Community Recovery and Mitigation	#15 - External Affairs
USDA			S		S	S		S		S	C/P	S		P	S
USDA/FS	S	S	S	C/P	S	S	S	S	S	S			S		
DOC	S	S	S	S	S		S		S	S	S	S	S	P/S	S
DOD	S	S	S	S	S	S	S	S	S	S	S	S	S	S	S
DOD/USACE			C/P	S	S	S		S	S	S	S	S	S	S	
ED					S										S
DOE	S		S		S		S	S		S	S	C/P	S	S	S
HHS			S		S	S		C/P	S	S	S			P/S	S
DHS	S	S	S		S	S	S	S	S	S	S	S	C/P/S	S	C
DHS EPR/FEMA		S	P	S	C/P	C/P			C/P	S				C/P	P
DHS/IAIP/NCS		C/P										S			
DHS/USCG	S		S	S				S	S	P			S		
HUD					S	S								P	S
DOI	S	S	S	S	S	S				S	P	S	S	S	S
DOJ	S				S	S		S	S	S	S		C/P/S		S
DOL			S		S	S	S	S	S	S	S	S		S	S

C = ESF coordinator
P = Primary agency
S = Support agency

Note: Unless a specific component of a department or agency is the ESF coordinator or a primary agency, it is not listed in this chart. Refer to the ESF Annexes for detailed support by each of these departments and agencies.

Figure 7.10 Matrix from the National Response Plan showing the 15 emergency support functions and agency assignments.

anism can be used to identify essential tasks and assign primary and supporting agencies to each. The success of this system depends on the support of the agency assigned to the task and planning products should be integrated into the strategic plan as performance objectives.

As has been noted, the precise list of functions used in preparing a functionally-based plan is the prerogative of the planners. There is no standardized list that is mandated for all plans, although SLG 101 is considered a best practice and constitutes the minimum level acceptable for emergency response plans. What makes the functional plan effective, however, is not the list of functions but the assignment of those functions to specific action agents. This assignment of responsibility carries with it the mandate to develop plans and procedures related to the function and to provide leadership for that function during disaster.

Functional planning correlates well with other areas of the emergency management program. It allows for the identification of those functions that will be needed during a crisis and a comparison of these functions to existing capabilities. The gap between needed and required capabilities can then be addressed through the strategic planning process.

CAPABILITIES-BASED PLANNING

Where scenario-based planning is designed around potential hazards and functional-based planning on the activities that will need to be performed in a crisis, capabilities-based planning (also sometimes called task-based planning) focuses on specific capabilities that will be needed in a crisis. In other words, it answers the question, "What capabilities will I need to perform my assigned functions?" These capabilities include:

- Personnel
- Planning
- Organization and leadership
- Equipment and systems
- Training
- Exercises, evaluations, and corrective actions

Capabilities-based planning is the system developed by the Department of Homeland Security for use in its national planning under Homeland Security Presidential Directive 8. DHS defines capabilities-based planning as "planning, under uncertainty, to provide capabilities suitable for a wide range of threats and hazards while working within an economic framework that necessitates prioritization and choice." It is an attempt to create an all-hazards capability while at the same time making more effective use of federal grant funds.

The DHS model is similar to the system that has been used very successfully by the military. The system breaks complex scenarios into a series of collective tasks and identifies the capabilities needed to meet the requirements of the given scenario. The model uses a series of related documents to identify "target levels of capability," that is the capabilities that will be needed, the level of government that will provide them, and the level to which they must be performed. These target levels of capability can then be compared with existing capabilities to determine shortfalls that can be addressed through existing programs or by new ones funded under federal grants.

The DHS model begins with scenario-based planning with the development of 15 National Planning Scenarios primarily focused on terrorist use of chemical, biological, radiological, nuclear and explosive (CBRNE) attacks (see Figure 7.7). The National Planning Scenarios were used to develop a Universal Task List that provides a comprehensive menu of some 16,000 tasks required to respond to the National Planning Scenarios. Besides its value as an operational tool, the Universal Task List provides a common terminology and reference system and can be used as the basis for training and exercise objectives. The tasks included pertain to all levels of government with no one level being able to perform all tasks. Planners select the tasks that apply to their level of planning (i.e. federal, state, or local government) and identify critical tasks that must be performed for operational success. Some 300 tasks in the Universal Task List have been designated as critical to operational success. A critical task is defined by DHS as one that must be performed during a major event to prevent occurrence, reduce loss of life or serious injuries, mitigate significant property damage, or is essential to the success of a homeland security mission.

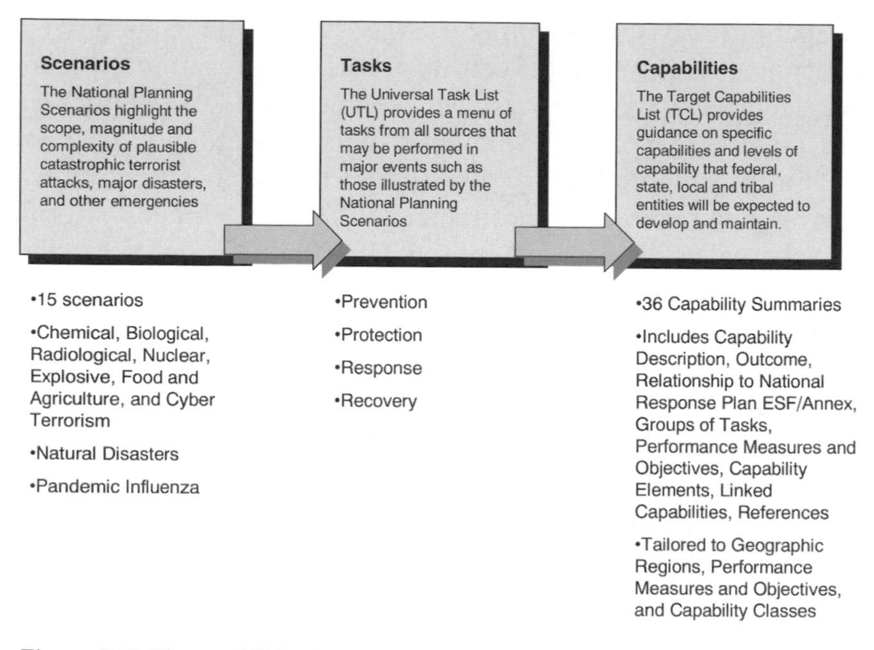

Figure 7.11 The capabilities-based planning model developed by the Department of Homeland Security.

The final document in the capabilities-based planning model is a Target Capabilities List. This list identifies 36 capabilities that DHS considers essential for performing the critical tasks identified in the Universal Task List. The expectation is that these capabilities should be developed and maintained, in varying degrees, by each level of government. The capabilities include performance measures, making the Target Capabilities List a potential tool for assessing performance. At the time of this writing, DHS is pilot testing an assessment tool based on the Woodbury Scale described in Chapter 4.

To a certain extent, capabilities-based planning combines elements of scenario- and functional-based planning and offers some advantages over using either method singly. The focus in capabilities-based planning is on capacity-development based on a realistic assessment of hazards, which is a basic goal of emergency management programs. Like functionally-based planning, the capabilities-based system seeks to build performance capacity, that is, the capacity to actually perform tasks required to meet operational objectives. Like scenario-based planning, the capabilities-based system uses projections of the most likely high-impact events.

There are some issues with the system, however. The most glaring flaw has been already mentioned—with the overwhelming emphasis on terrorism, it is unclear whether all the tasks and capabilities needed for a true multi-hazard approach have in fact been identified. Further, with over 1,600 items in the Universal Task List and with the Target Capabilities List (TCL) numbering over 170 pages, accessing the information in the documents is not easy and developing plans based on the system can be cumbersome. Despite the length and complexity, the documents are extremely generic and provide little real guidance to planners.

Another concern is the lack of correlation between the DHS model and existing standards such as EMAP and NFPA 1600 and accepted best practices such as SLG 101. The DHS model recognizes four mission areas: prevent, protect, respond, and recover, a departure from the four-phase Comprehensive Emergency Management model that forms the basis for most emergency management programs in the U.S. While the TCL identifies capabilities for post-disaster mitigation and economic recovery, it does not scratch the surface of complex recovery issues. This makes the program too easy to dismiss as "just another terrorism plan" and will no doubt generate resistance to its implementation at the local level. Because grant funding will no doubt drive compliance with the TCL, there is also the possibility that local emergency management offices will focus solely on the TCL capabilities and neglect the critical functions of mitigation, continuity, and recovery because they are not required by the federal government.

There may also be similar problems in the use of the TCL as an assessment tool. DHS has expressed concerns that EMAP does not assess capability and hence is not useful as a measure of performance. This is an "apples and oranges" argument. NFPA 1600, the standard that forms the basis of EMAP was established to assess programs against a common set of criteria. It encourages capacity development by defining program outputs but was never intended to include specific measurements for assessing capability. Since the EMAP program is voluntary, encouragement by the federal government for jurisdictions to adopt the standard is essential to its success. There is concern that DHS is adopting an "either or" mindset that could lead to the abandonment of EMAP. In reality, EMAP and NFPA 1600 provide the strategic context for planning, while the assessment of capability envisioned for the TCL could provide a useful tool for program evaluation. The two together would be a powerful combination for improving crisis response.

Finally, the DHS model represents precisely the type of detailed planning that social science suggests does not work well in coordinating disaster response. In this approach, one finds the type of mindset and narrowness of vision warned of by Weick and Suttcliffe. Inherent in the model are the assumptions that the next event will most likely be a terrorist attack, that it will require strong, centralized command and control, and if all the target capabilities are achieved, success in response is assured. There is no provision in this type of model for the improvisation and creative problem solving that characterizes successful disaster management.

In recent years, DHS has also shown a bias in favor of many of the complex systems used by the military. However, the military is a hierarchical system that is, to a certain extent, very homogenous. The complexity of civilian disaster response with its competing special interest groups and multiple layers of coordination requires a cooperative rather than a hierarchical approach, as was discussed in Chapter 2. The critical collective tasks required of civilian disaster responders do not lend themselves as readily to a checklist format as do military tasks. Whether such a system will be successful in a less hierarchical environment is still to be determined.

GENERAL PLANNING PRINCIPLES

Over the past several chapters, this book has considered a number of concepts, principles and practices related to the development of emergency plans. Taken together, they constitute a comprehensive set of principles that can be used to assess and improve community disaster planning. These principles are not new—they are from the work of Dr. E. L. Quarantelli and were first articulated as early as 1991. These principles are:

1. Focus on the planning process, not the plan. As was noted above, a written plan reflects the intent of the community as to how it will deal with crisis—it is not an end in itself. Planning is about bringing community elements together under a shared vision and the process of achieving this consensus is more important than the final end product.
2. Planners must recognize the qualitative and quantitative differences between emergencies, disasters, and catastrophes. As was discussed in Chapter 2, the differences between these three categories of crisis are significant and important to the scope of the plan.
3. Be generic rather than agent-specific. This principle is the heart of the functional planning methodology described above. It assumes a commonality of response to various crises that is independent of the triggering agent. This allows true all-hazards planning.
4. Avoid development of a "command and control" model. As was discussed in Chapter 2, authoritarian models based on military systems are not generally effective in disasters, despite the recent emphasis on the National Incident Management System. This is because such systems are based on a misunderstanding of how people actually react in disasters and attempt to impose an artificial structure on behavior that is normal and predictable. Instead of direction, response models must focus on coordination.
5. Focus on general principles rather than specific details. A common practice in plans is to become too detailed, to attempt to cover too much information. As was noted above, plans should be written to the level at which they will be used. Plans are guidelines that should assist in the creative problem solving that characterizes successful response.
6. Base plans on what is likely to happen. A common practice is to respond to the last disaster. The actions of the federal government

since September 11 are a classic example of this tendency, with Hurricane Katrina serving as the case study of what can occur if one constantly looks behind rather than ahead. A realistic appraisal of potential hazards is the foundation of all planning.

7. Plans should be vertically and horizontally integrated. Planning must be an inclusive process. A characteristic of disasters is a convergence of responders from both the public and private sectors and from within and outside the community. The more these disparate elements are integrated into the planning process, the greater the chance for successful response.
8. Anticipate problems and suggest solutions or options. A characteristic of crisis is the need for creative problem solving. There is an old military maxim that states, "No plan survives contact with the enemy." The same is true during a crisis. Good planning should provide a vehicle for coordinating response and help provoke appropriate actions by offering options and potential solutions that can be considered within the operational context of the crisis.
9. Use the best social science knowledge possible. Chapter 2 discussed many of the common disaster myths and their negative influence on planning. It is startling how many emergency managers are unaware of information about the nature of disaster that has been available in social science research for years. By neglecting social science research, emergency managers are condemned to consistently repeat the same mistakes.
10. Recognize that there is a difference between pre-disaster planning and planning done during a disaster. In a sense, all plans are strategic in that they represent what the community expects to happen and how the community expects to respond. However, once the disaster begins, plans are overcome by events and creative problem solving and tactical response are based on the realities of the situation. The process of developing plans prior to disaster and the process of planning during a disaster are separate processes.

Quarantelli's ten principles of disaster planning represent a fusion of social science and accepted emergency management principles that create a standard for community planning. They also suggest why so many emergency plans seem to fail when implemented during crisis. Failure can usually be traced back to a violation of one of these principles. Of more concern is the implication that emergency planning in the United States has been skewed toward an ineffective model for over 50 years and that it is likely to get worse with the increased emphasis on terrorism.

CONCLUSION

Planning is the logical outgrowth of strategy development. It is the mechanism used to translate strategic policies and concepts into practice and to identify the resources that will be needed to respond to crisis. However, it is the process of planning that is important rather than written plans. Once this is understood, the importance of engaging other groups in the process and the need to be inclusive becomes clear.

An important concept is the idea that plans are not always implemented sequentially and that the activation of multiple plans can lead to a competition for limited resources if they have not been synchronized through a coordinated process. This suggests that the Comprehensive Emergency Management Model is a strategic concept rather than a tactical one and that plans are better considered based on the time of implementation. Instead of considering sequential response, continuity, and recovery actions, one approaches holistic planning from the point of view of what tasks begin immediately, which are mid-range actions, and which are long-range. Under this approach, an emergency operations plan may include elements related to life-safety, continuity, recovery, and post-disaster mitigation.

Effective planning requires an understanding of the nature of disaster, the risks facing the community, and the availability of community resources. This helps to ensure that plans are based in reality and do not represent just wishful thinking on the part of the jurisdiction. To this end, it is important to use a structured process such as the problem-solving model and to synchronize various plans to create a holistic community response.

Chapter 8

PLANNING TECHNIQUES AND METHODS

The intelligence quotient of any meeting can be determined by starting with 100 and subtracting five points for each participant.
—Scott Adams

The previous chapter considered basic concepts that provide a context to plan development. But how does one actually make progress in developing plans? Plans, particularly those involving multiple jurisdictions or agencies, seem to take forever to produce. There are endless meetings of well-intentioned representatives and numerous draft plans and proposals, but final products seem elusive. This chapter looks at some of the techniques and methodologies that can be used to develop a productive plan.

ESTABLISH A PLANNING STRUCTURE

A critical first step in developing plans is to determine up front who are the stakeholders and what mechanism will be used to gain final approval for the plan. Stakeholders are those who have a vested interest in the plan, either through responsibility for implementing the plan or by virtue of being affected by the plan's implementation. This could conceivably be a large group and, as Scott Adam's humorous quote cautions, involving too many players at a meeting can hamper progress. Consequently, it is necessary to determine which stakeholders play crucial roles in the plan as decision makers and implementers versus those stakeholders who do not have a direct role but whose needs and support must be considered in plan development. This allows the establishment of a planning structure that limits the number of final decision makers yet provides for the inclusion of all stakeholder concerns during plan development.

As part of the process of identifying stakeholders, thought should be given to who will actually "own" the plan. Owning the plan means to accept responsibility for overseeing plan development and maintenance. Normally, this is the department or agency that has the lead in implementing the plan. However, this responsibility, especially where multiple departments or agencies are involved, is not always clear. Identifying this key agent becomes particularly important in plan maintenance. Plans are a snapshot of current thinking by the planning body, but the planning process continues past the publication of the plan. Over time, assumptions may change, authorities may shift, and lessons are learned through exercises and actual experience. There will be a need to update the plan, either in whole or in part, and it is essential that someone be responsible for monitoring changing circumstances and coordinating plan changes.

Chapter 4 discussed the role of the advisory committee required under NFPA 1600 and introduced the FIRESCOPE planning model as one possible method for organizing the emergency management program. In this model, the work of specialist teams or inter-agency task forces was coordinated by an operations team and approved by the advisory committee. This approach has the advantage of having planning done by those who will be responsible for implementing the plan while still allowing for oversight of the products of the team. It incorporates the deference to expertise that Weick and Suttcliffe note is one of the characteristics of high reliability organizations.

Another model for planning is shown in Figure 8.1. This model uses a smaller steering committee to coordinate the work of various working groups drawn from the larger group of stakeholders. The steering committee meets fairly frequently to monitor progress and to develop the agenda for the next group meeting. The larger group would meet less frequently to review and comment on projects done by the work-groups. Completed projects are then submitted for approval to the advisory committee. Both the FIRESCOPE and steering committee models provide for the inclusion of stakeholders in the planning process, the assignment of work, and the final approval of the work products.

This approval of work products is a critical step in planning. When operational experts are primarily responsible for planning, there is risk that the end product will be written to the operational level. This

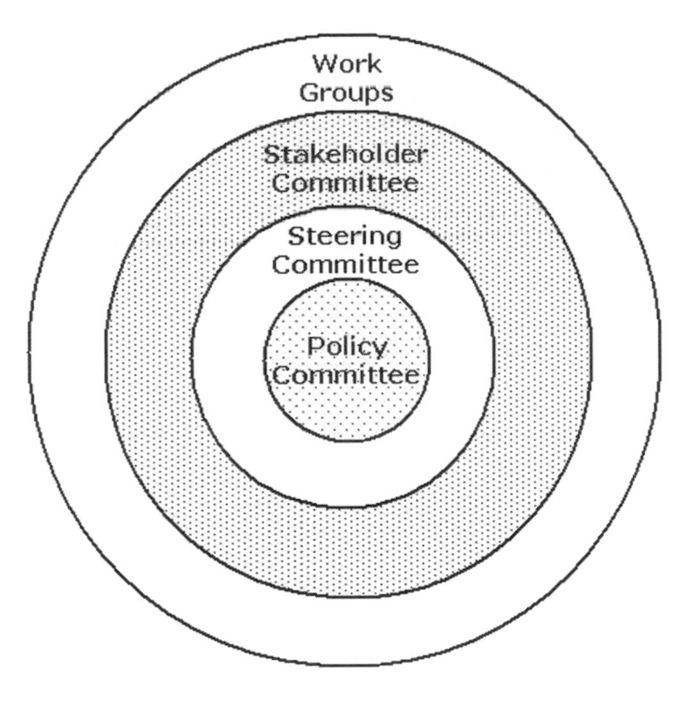

Figure 8.1 Emergency planning model.

means that the scope of the plan will be narrow and it may not be fully integrated with stakeholder agencies that do not have direct operational responsibility. If the aim is to produce a tactical plan that will serve as a coordination mechanism for other plans, this could present a problem. Consequently, part of the governance structure for planning should include the provision for oversight of the end product, to include the circulation of draft documents to all stakeholders before final decisions are made.

In establishing the planning governance structure, thought should be given as to how the group will make decisions. Conflicts can be decided in several ways. Two of the most common in a group setting are executive decision and majority vote. In the executive decision model, a senior manager agreed upon by the group is the final arbiter of any conflict within the group and his or her decision is considered final. In the majority vote model, members of the group who have voting privileges settle conflicts by voting with a simple majority or an agreed upon ratio (e.g. two-thirds majority) deciding the outcome.

The problem with these two common methods is that they constitute what psychologists call a win/lose scenario. In this transaction, someone has won (the side that the decision or vote favored) and someone has lost and is consequently unhappy. There can also be issues over who can vote. For example, are all representatives from a department allowed a vote or only one? Do private companies or volunteer groups have a vote? For these reasons, a more reasonable approach is to resolve conflict through developing a consensus. Consensus means that everyone in the group can live with the decision without feeling that they are surrendering anything significant.

Consensus is achieved by a process of discussion to determine the elements of disagreement and to resolve them by either providing more detailed information on the issue or by finding a way of achieving the same result through a slightly different approach. A useful starting point is to focus on what the group already agrees upon before tackling the areas of conflict. This establishes a basis for cooperation and puts the points of conflict in context. Consensus can generally be achieved when group members are able to articulate their concerns. These concerns are frequently based on assumptions about other group members' positions or information that the group had not considered. Once the concerns are articulated and examined by the group, it is usually possible to resolve points of conflict in a manner that allows the group to move on.

USE A MEETING FACILITATION PROCESS

Achieving consensus depends on creating a safe environment for free discussion of points of conflict and establishing a process that leads to resolution of those issues. To create this environment and process requires that meetings be carefully structured to achieve desired outcomes. The meeting is an essential tool of the emergency planner. An effective plan requires the support of multiple stakeholders and the best way of garnering that support is to involve stakeholders in the planning process. Consequently, at some point these stakeholders must be brought together in a meeting.

Unfortunately, most meetings tend to be unfocused and counterproductive. This is usually the result of differing expectations among attendees and a lack of agreement on the purpose of the meeting. Decisions made at the meeting are not captured for those who were not present and there is seldom any follow-up or action taken outside the meeting. In many cases, the meeting is viewed as an end in itself rather than part of a process that leads to the development of a work product. The result is normally a seemingly endless series of meetings that produces few results.

To counteract this tendency for meetings to degenerate into nonproductive gatherings, it is necessary to use a formal process to facilitate the meeting. There are a number of different methods for meeting facilitation, such as the Interaction Method pioneered by Interaction Associates of San Francisco. Such a process both keeps the meeting focused and allows for assigning and tracking follow-up actions to make sure that progress is in fact being made. A meeting facilitation process has three core components: preparation, conduct, and follow-through.

A frequently overlooked meeting tool is the agenda. The agenda is the mechanism that helps form a common expectation for the meeting. It is derived through a process of determining what the issues are and how the meeting is expected to address them. This process determines what issues will be excluded, how issues will be presented, and how

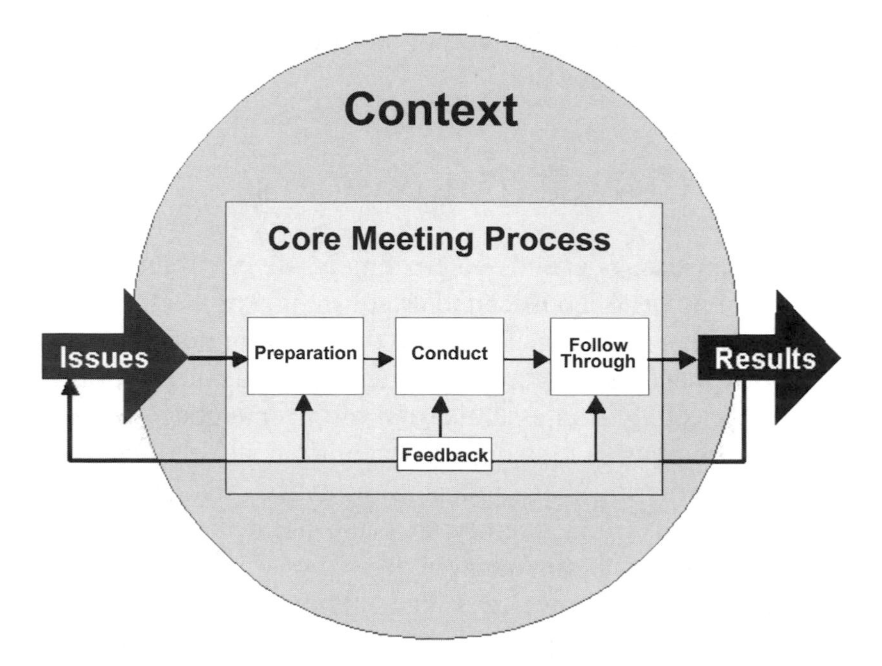

Figure 8.2 Meeting facilitation process. Based on a diagram by Interaction Associates© 1986.

decisions will be made (e.g. consensus, majority vote, executive decision, etc). This allows attendees to make decisions about participation in the meeting. For example, an attendee may decide to bring a senior representative, subject matter expert, or appropriate files and documents depending on the issues under discussion. In the worst case, an attendee may decide not to attend in order to deal with a higher priority. While this may seem a problem at first, it is merely good time management. Wasting attendees' time is a cardinal sin in meeting management.

The agenda identifies:

- Expected outcomes for the meeting—effective meetings produce results. The end result of the meeting should be clearly articulated so that every attendee has the same expectation. Knowing the expected outcome can have a bearing on the level of attendance. For example, if the expected outcome is to make a final decision on an issue, a senior official may elect to attend the meeting rather than sending a representative.

- Topics to be discussed—knowing what topics will be discussed allows attendees to properly prepare for the meeting, either through conducting background research or by bringing appropriate materials. It also allows busy officials to determine whether the meeting is appropriate for them to attend.
- Group Leader—someone must accept responsibility for calling the meeting. The perceived importance of this group leader may also affect the level of the attendance.
- Process for each agenda item (formal presentation, group discussion, decision, etc.)—this item informs the attendee of what to expect for the topic to be discussed. This prevents misunderstandings about whether issues will be discussed or decided on and again helps the attendee make decisions about attendance and pre-meeting preparation.
- Time limits for agenda items—effective meetings start and end on time. This is a key point—busy officials willing to commit time to a meeting must be assured that the meeting will not run over its scheduled limit. The way to manage this is to set time limits for each agenda item and stick to them. Attendees should know this in advance.
- Meeting information (e.g. time, date, location)—while it may seem obvious, an agenda should contain basic information regarding the meeting time, date, location, parking arrangements, etc.

The second core component of the meeting facilitation process is the conduct of the meeting itself. The facilitator is concerned about several key processes: encouraging participation to the maximum extent possible, capturing key discussions and decisions, and keeping the meeting focused on the agenda and expected outcomes. To accomplish this, the facilitator has a number of tools available:

- The Agenda—the first action in every meeting should be to review and agree on the agenda. If necessary, the group can mod-

Agenda	Strategic Planning Committee 8/16/200X 1:00 PM to 2:00 PM Building 9
Meeting called by:	CCCC, Chair
Type of meeting:	Planning
Facilitator:	XXXX
Expected Outcomes:	1. Approve draft Administrative Plan 2. List of proposed goals and objectives

Order of Agenda Items	Persons Responsible	Process	Time Allocated
Agenda Review	XXXX	Present	5
Approve Administrative Plan	Committee	Discuss Decide	10
Review proposed strategic goals and objectives	Committee	Present Discuss	35
Next Steps	Committee	Discuss Decide	5
Meeting Review	XXXX	Discuss	5

Figure 8.3 Sample agenda format. The agenda helps manage expectations for the meeting.

ify the agenda. However, once the agenda has been agreed upon, it should be followed and any additional items or extended discussions should be placed on a future agenda.

- Ground Rules—although they may appear unnecessary, ground rules can prevent potential conflict and irritants. They establish a safe environment and limit interruptions. Examples include prohibitions on personal attacks, agreement to stick to time limits, and placing cell phones and pagers in vibrate mode.
- Group Memory—a key role in the meeting is that of the recorder. However, just having someone taking notes does not always sat-

isfy all meeting participants, particularly someone whose principal concern is not on the agenda. A flip chart or electronic projection system allows participants to see that an item or issue has been recorded using language that captures the essence of the discussion. It legitimizes participants' statements, provides accountability, and frees participants from note taking. The group memory also forms the basis of the minutes prepared after the meeting. A section of the group memory should be designated to record items that are unrelated to the agenda but that the group wishes to consider at a future meeting. This "parking lot" allows items to be captured for later consideration without compromising the agenda.

- Action Plan—the action plan captures next steps for the group. It identifies the action to be accomplished, the action agent, and the timeframe allotted for the action. This is an essential component to the meeting if progress is to be made.
- Meeting Review—it is useful to conduct a brief review as the last agenda item of the meeting. This review focuses on what the participants liked about the meeting and what actions could be taken to improve future meetings.

Oddly enough, very few groups produce a written record of what was accomplished at the meeting. This is normally because of the time required to develop formal minutes or through some misguided notion that the minutes may be used against the group in litigation. Yet keeping a record and tracking progress are critical to making meetings productive. It allows stakeholders who were not at the meeting to be kept apprised of key discussion points and decisions and it allows the group to measure progress. By using a facilitation process, the task of producing a written record is not as onerous as producing formal minutes. This is because several key components of the written record are produced at the meeting and can be included as part of a meeting memorandum in lieu of formal minutes. The basic components of the meeting memorandum are:

- Agenda—the meeting agenda is included in the meeting memorandum to show the items under discussion.
- Summary of key decisions—this summarizes any decisions made by the group in accordance with the agenda.

- Group memory—the group memory is recorded directly as written.
- Unresolved issues—these are issues that are potential agenda items for future meetings. They are drawn from the "parking lot."
- Action plan—the action plan is usually given as a table showing the action item, the action agent, and the target date for completion.
- Meeting review—usually reflected in two columns: went well/could be improved.

By using a meeting facilitation process, meetings become more productive and participants can see progress being made toward the group's objectives. The sidebar case study from Hurricane Iniki

Meeting Memo	**Strategic Planning Committee** **8/16/2005** **1:00 PM to 2:00 PM** **Building 9**	
Meeting called by:		
Type of meeting:		
Facilitator:		
Expected Outcomes:		
Present		
Invited		
What Happened and How:		
Decisions/Action Items:		
Decisions: Action Items:		
Item	Who	When
Next Steps:		

This meeting memo is my interpretation of what happened at this meeting. If you would like to correct an error or make an addition, please contact XXXX.

Signature

Figure 8.4 Sample meeting memorandum. This document is quicker to produce than standard meeting minutes and provides all essential information about decisions made and actions assigned.

demonstrates the simplicity and power of the meeting facilitation process. In this case, the sheer volume of information and the many unrelated activities that needed to be coordinated overwhelmed the federal responders until the implementation of a facilitated meeting process.

DEVELOP AN ACTION PLAN AND SET DEADLINES

Another frequent occurrence in planning is the establishment of an open-ended task force or working group. In reaction to an occurrence or recognized need, a group of experts is brought together to "do something" about the problem. After an enthusiastic initial meeting, the group continues to meet without producing anything until it disintegrates from lack of interest. There is usually a fair bit of reluctance to join such groups initially and participants, even though supportive of the group's purpose, will eventually come to view the meetings as a waste of valuable time and personnel resources.

A simple way of approaching such a problem is to develop an action plan up front that defines the expected outcome and establishes milestones to be accomplished at each of a set number of meetings. Using this methodology, participants are asked to commit to a specified amount of time and develop a defined deliverable. The action plan can be modified if the group determines more time is needed. Since it is already agreed that the group will have a limited tenure, these time extensions should be manageable.

Action plans are relatively simple in concept. However, a major stumbling block is usually gaining agreement on the deliverable. There is usually a tendency for group members to attempt to add pet projects, resulting in what the U.S. military terms "mission creep." It is critical, therefore, that the initial meeting be used to develop agreement on the scope of the project and the end product. Once these two issues are resolved, the group can identify tasks that will need to be accomplished to achieve the end state and can assign action agents to those tasks.

The establishment of a project timeline also allows for management oversight of the working group. The group leader can now be held accountable for meeting project milestones and has the opportunity to

CASE STUDY: HURRICANE INIKI, 1992— MEETING FACILITATION AIDS RESPONSE

Case Figure 8.1 Hurricane Iniki struck the Hawaiian Islands on September 11, 1992. It was the worst storm in recorded history to strike the islands, and one of the costliest hurricanes in U.S. history. Courtesy of the National Oceanic and Atmospheric Administration.

On September 11, 1992, the eye of Hurricane Iniki passed over the Hawaiian island of Kauai, causing six deaths and over $3 billion in damages. Over 14,000 homes were damaged or destroyed and electric power and telephone services were completely disrupted. Damage to cash crops of sugar cane, papaya, and banana and to the local hotel industry caused severe economic hardship. It was the worst storm in recorded history to strike the Hawaiian Islands and, at the time, one of the costliest hurricanes in U.S. history.

Almost before the storm had passed, the Federal Emergency Management Agency began deploying federal resources to assist in relief and recovery operations. Because of the extensive damage to Kauai, FEMA implemented the fledging Federal Response Plan, deploying over 20 federal agencies and the Red Cross. A management team from FEMA Region IX in San Francisco arrived on Oahu on September 12 and began operations from a temporary facility.

(continued on next page)

(continued from previous page)

The management team was immediately overwhelmed by the data being provided by the other agencies. Just reading and collating situation reports from federal and state agencies swamped the small plans unit. The operations section faired little better—so many resources were being deployed that daily operations meetings took several hours and resulted in a mass of data that had to be reviewed and collated. The operations chief was unable to give an acceptable daily briefing to senior management because of this volume of information and critical issues were frequently overlooked.

In desperation, the Plans Chief sought help from the U.S. Forest Service personnel who had been deployed to deal with potential fire fighting issues. The USFS had deployed a trained incident management team that was not being used and provided personnel from its plans unit to assist the FEMA team.

One of the first things that the USFS personnel recommended was the use of a facilitator at the daily planning meetings. They established an agenda that focused on unresolved critical issues rather than on a "brain dump" and formulated a set of ground rules to guide the meeting. The Operations Chief was initially skeptical, fearing that he would not have all the information he needed to coordinate activities. However, in a few days he began to realize that he was now fully aware of problem areas and could formulate action plans to deal with them. The meeting minutes formed the nucleus of his daily operations briefing to senior management and he was able to present significant issues in a focused and efficient manner. Operations meetings went from three hours to a half hour and personnel left feeling much more comfortable with the progress of the operation.

The use of a facilitated meeting process shortened the length of the meetings, made the time spent in meetings more productive, and focused staff on essential requirements. The results were so successful that FEMA adopted the USFS methods as part of the standard operating procedures for FEMA plans units.

explain any lack of progress to a group that can assist in providing additional resources or removing stumbling blocks to achieving the project's goal.

Managing Multiple Projects

While the development of action plans and project timelines help manage an individual project, it is readily apparent that emergency planners must manage many projects at the same time. These multiple demands make it difficult to focus on important tasks and can rapidly diffuse the effectiveness of an emergency management office. Projects are not completed, staff become exhausted and frustrated, and there is a general impression of lack of progress. The stress of overseeing an emergency management program can never be eliminated, but can be lessened by organizing the office to allow efficient management of projects.

The principle tool for managing multiple projects is the annual work plan. To a certain extent, the work plan fulfills a role analogous to the strategic plan. Where the strategic plan identifies goals, performance measures and resources for the program as a whole, the work plan performs a similar task for the emergency management office and staff. Work plan elements are essentially individual performance objectives and are crafted in much the same way as those in the strategic plan: they identify the task, assign an action agent, establish deadlines, provide a metric, and provide resources.

Figure 8.5 shows an example of a work plan element. Note that it identifies a specific goal that is then broken into measurable tasks, each with an assigned action agent, a completion date, and an expected outcome. There is also an estimate of the time required for the project. In this case, the staff member Jones is the lead on the project and is expecting to spend over half of his or her available time on this project. This estimate becomes particularly important when one summarizes the total work to be performed by an individual. If the total work assigned exceeds available hours, then tasks will need to be reprioritized or other resources such as additional staff, contractor support, or an overtime budget must be provided. This process keeps staff from being overloaded.

Projects can also be prioritized within the work plan. A common method of ranking is:

- Priority 1—must be accomplished
- Priority 2—should be accomplished
- Priority 3—would like accomplished

Goal 5 Revise shelter operations plan					
Task	**Assigned To**	**Approximate FTE**	**Estimated Completion**	**Measurement**	**Budget**
5.1Conduct survey of existing shelters using contractor support	Smith	.2	3/10	Updated shelter database	$15,000
5.2 Convene shelter working group and develop revised plan	Jones Johnson	.5 .2	9/1	Draft shelter operations plan	N/A
5.3 Conduct tabletop exercise to test revised plan	Jones	.1	9/15	Exercise after action report	N/A
5.4 Submit revised plan to Advisory Committee for approval	Jones Director	N/A	11/5	Revision approved	N/A

Figure 8.5 Sample performance elements from a work plan.

In actual practice, however, these priorities are not particularly effective. Priority 1 tasks are generally accomplished, Priority 2 tasks may get partially accomplished, but Priority 3 tasks are not even considered because of lack of time. Realistically estimating the time required for a project or task and prioritizing these on the basis of available resources is a much more effective management tool. It also has the bonus of being defensible. One does not have to explain why tasks identified in the work plan as Priority 2 or 3 were not accomplished.

Another key element in the work plan is the measurement of success. If one expects the emergency management program to produce results, one should be able to identify the expected outcome in measurable terms. In this example, each performance objective has a documented outcome (a database, a plan, or a report) or a specific verifiable action (advisory group approval that is documented in meeting minutes). This measurement has two purposes: it allows the planner to understand the emergency manager's expectation for the task and it provides a metric for the manager to verify that the work has been accomplished.

Like the strategic plan, the work plan is not a static document. Although prepared annually, it should be reviewed quarterly or semiannually and adjusted as necessary. Adjustments may be necessary because projects may be delayed, staff changes may occur, or new priorities may arise. The work plan is a tool to manage the activities of the emergency management staff—it is not intended to lock the emergency manager into a pre-determined course of action.

Another tool that may be useful in managing workflow is the use of graphic planning models. These models provide visual representations of the work plan and help to identify available time or critical paths. While not generally used by emergency management offices, they do offer some useful advantages in managing staff time.

The simplest graphic model is the Gantt chart. Originally developed by Henry Gantt in 1910 to track manufacturing processes, the Gantt chart has developed into a standard management tool. The strength of the chart is that it tracks project initiation and completion over time and provides a visual representation of project commitments. Figure 8.6 is an example of a Gantt chart developed using the work plan elements from Figure 8.5. In this case, there is a timeline for the entire project and individual timelines for each task. The same process could be used to track an individual employee's work assignments and, using an expanded calendar, factor in non-available time such as vacations, exercises, mandatory meetings, etc.

While a graphic tool such as the Gantt chart and its more complicated counterpart, the Program Evaluation and Review Technique (PERT) chart, can be extremely useful tools, they are not a substitute for the development of a structured work plan. Without the specific performance objectives delineated in the work plan to provide a basis for charting, graphic models have only limited use in managing workflow.

FACILITATE DECISION MAKING

An emergency planner must be something of a psychologist in terms of understanding human nature. Decision making is difficult for many people because of concerns about the impact on the decision maker if he or she makes the wrong call. This is usually the result of the decision maker feeling that he or she does not have sufficient information to make an informed decision. Similarly, many decision makers cannot articulate the outcome that they desire from a project beyond very gen-

	JAN	FEB	MAR	APR	MAY	JUN	JUL	AUG	SEP	OCT	NOV	DEC
Goal 5												
Tasks												
5.1												
5.2												
5.3												
5.4												

Figure 8.6 Sample of the use of a Gantt chart to display work plan information. This chart is derived from the performance objectives in Figure 8.5.

eral terms. Understanding this, the emergency planner can structure the planning process to overcome these two psychological barriers.

A key to overcoming these barriers is to understand that senior officials make decisions on the basis of trust in the process and the project leadership, not in a detailed reading of the final end product. In other words, a senior official will rarely read an entire plan but will instead approve or disapprove it on the basis of a briefing on the plan, the process used to develop it, and the results of reviews by trusted advisors. A plan that is the result of a collaborative multi-agency project and approved by intermediate managers will most likely receive the approval of senior management. The senior official assumes that key personnel with expert qualifications have already reviewed and approved the plan and that approval by the senior official carries low risk.

Once this process is understood, planning groups are better able to make decisions regarding the planning project. It is not uncommon for planning projects to be delayed while awaiting decisions on policy issues that surfaced during planning. These issues are usually framed as a question, rather than as a recommendation. In essence, the planners are asking a senior official to make a decision for which the official has little or no knowledge or experience. The dynamic changes if planners realize that as subject matter experts their opinion carries weight with senior officials. Instead of asking at each step, "What do you want me to do?" The planners now can say, "Here's my best recommendation." Planning is at its most effective when it is bottom-driven in this way.

A critical starting point in the planning process is to define expected outcomes and planning parameters that affect these outcomes. While one would expect that the senior executive would define these, in actual practice it is common for the emergency planner, working with other managers and the planning group, to develop recommendations for consideration by executive management. There are policy level decisions that have to be made before the rest of the planning process can continue. These policy decisions will form the basis of all future plans. Again, approaching senior management with a recommendation rather than a question simplifies the decision making process. The senior executive can approve or disapprove or request more information but he or she does not have to synthesize ideas with too little information.

A useful technique for obtaining a senior management decision is the decision paper. A decision paper lays out the thought processes behind a policy recommendation and its potential impact on the organization. Figure 8.7 shows a decision paper format used by the United States Army. This particular format allows for a brief description of the reason why the decision is needed and shows how the recommendation was developed. It also provides an analysis of the impact of the decision (e.g. costs, personnel requirements) and provides an opportunity for review by affected stakeholders. The latter helps diffuse potential conflict by allowing a stakeholder to disagree with the recommendation while providing opportunity for the planner to rebut their disagreement. Using a formal mechanism such as the decision paper provides the senior executive with all relevant information needed to make a decision and significantly simplifies the process. A simple approval signature on the document is usually sufficient to implement the policy.

Decision making within working groups can also be facilitated through a similar process. It is difficult for large groups to actually

1. **Problem** – a brief statement of the problem that led to the policy recommendation
2. **Recommendation** – the policy recommendation
3. **Background and Discussion** – Facts relevant to the problem such as other solutions that were considered and their pros and cons
4. **Resource Impact** – Potential impact to the organization such as budget costs, redirection of resources, etc.
5. **Coordination** – Agreement or non-agreement by stakeholders who will be affected if the policy is implemented
6. **Consideration of Non-concurrence** – Rebuttal arguments to any non-agreement by stakeholders.

Figure 8.7 Decision paper format. Providing a recommendation rather than a question helps senior executives make decisions quickly.

develop a work product while at a planning meeting, so it is common for several individuals to develop draft products and present them to the group for comment. A quick way of jump-starting a group is to prepare what the military calls a "straw man," a draft prepared and distributed in advance of the meeting that will provoke discussion. The concept of the straw man is based on the theory that it is always easier to criticize an existing document than to synthesize a new one. Where time is important, collecting assumptions, proposals and ideas into a draft can provide a working document, or, at the very least, generate discussion over points of contention.

Like all components of a project, reviews of draft documents should have deadlines. This deadline can be a meeting to discuss the draft or a date by which comments must be submitted. Ideally, stakeholders should either comment or signify agreement in writing. Less desirable but also acceptable is to have a group agreement that if no comments are received from a stakeholder by the deadline, it is assumed that they concur with the draft. In fast paced situations, it may be possible to use a deadline and a return receipt as proof that the stakeholder received the document and had an opportunity to review it, even if they chose not to comment.

USE COMMON PLAN FORMATS

For all the diversity among emergency management programs, there is a remarkable similarity in emergency plan formats. Interestingly enough, this has been the result of misinterpretation of federal guidance on emergency planning.

FEMA's first major foray into emergency planning guidance was Civil Preparedness Guide (CPG) 1-8 *Guide for the Development of State and Local Emergency Operations Plans* and its companion document CPG 1-8A *Guide for the Review of State and Local Emergency Operations Plans*. CPG 1-8 established a common format for emergency plans and CPG 1-8A was a crosswalk for evaluating plans to make sure all key components were included. As requirement for participation in the State and Local Emergency Management Assistance (EMA) grant program, state and local governments were required by 44CFR to "conform to the requirements for plan content" established in these documents.

Despite the requirement for conformance and not compliance, emergency plans began to use the sample emergency operations plan shown in Chapter 6 of CPG 1-8 as the basis for all planning. State and Local Guide (SLG) 101 *Guide for All-Hazard Emergency Operations Planning* superseded CPG 1-8 in 1996. SLG 101 did not make any significant changes to the emergency operations plan format and was specific that the format was not mandatory (see Chapter 3 of SLG 101). Nevertheless, within a few years, it was difficult to find a jurisdiction that did not follow the SLG 101 sample format.

Interestingly enough, when FEMA developed the Federal Response Plan, it chose a format that, while conformant with the principles of SLG 101, did not use the SLG 101 sample format. Instead, FEMA opted to emphasize the functional planning approach by grouping federal responders into a series of Emergency Support Functions (ESF),

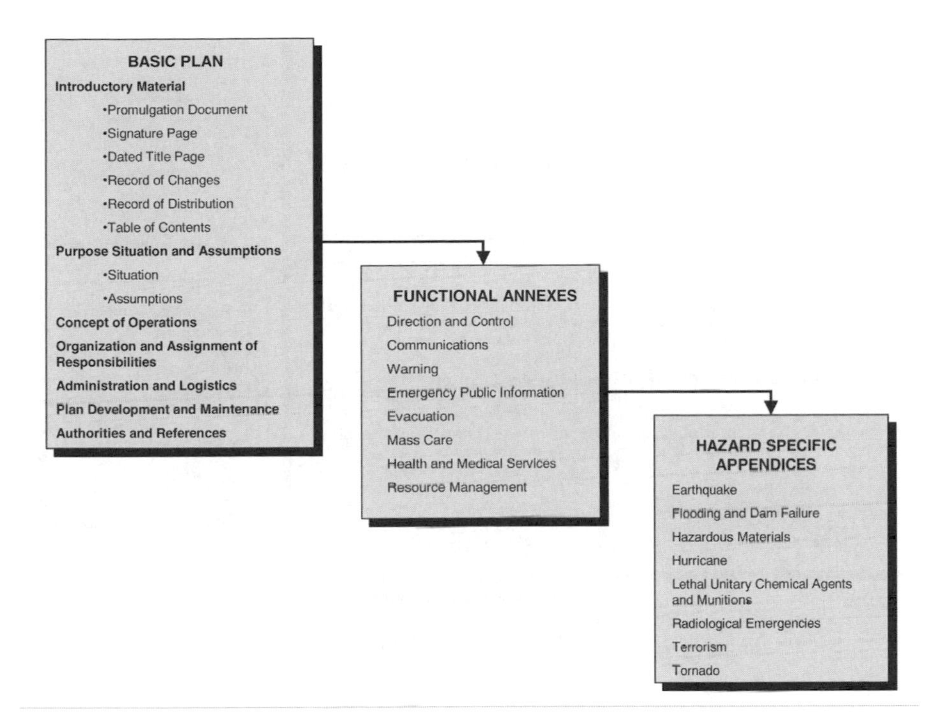

Figure 8.8 Emergency Operations Plan format from SLG 101 *Guide for All-Hazard Emergency Operations Planning.*

each with its own functional annex, and added a series of support annexes for those functions not assigned to an ESF. The plan also included an incident specific annex and included plan information and maintenance in a series of appendices. This same format has been carried over into the current National Response Plan.

With the issuance of the National Response Plan, emergency operations plans are undergoing another revision as states scramble to realign their plans to the one used by the NRP. This change is taking place under the misapprehension that the National Incident Management System and other DHS initiatives require that emer-

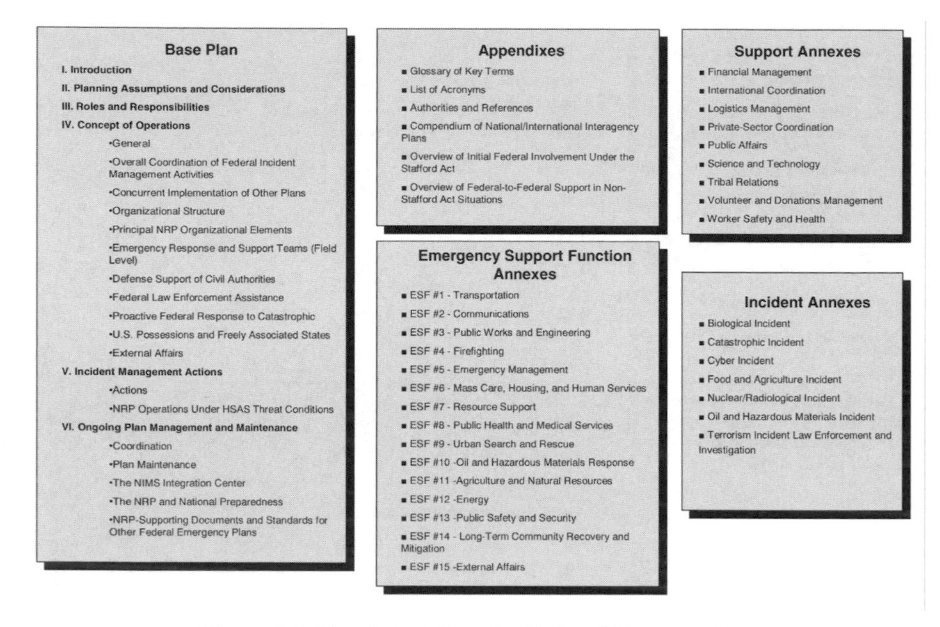

Figure 8.9 Plan format from the National Response Plan.

gency plans align with the NRP. Not only is there no such requirement, at the time of this writing SLG 101 is under revision to align it with NIMS and is expected to be fielded by 2007.

This misunderstanding of conformance versus compliance is one of the reasons that the EMAP Standard and NFPA 1600 do not include sample plans. There are concerns that any samples would quickly be adopted as requirements for demonstrating conformance with the standards. Nevertheless, using common formats for various plans offers several advantages:

- Information is easier to locate in a crisis if similar formats are followed.
- Assessing plans for conformance with jurisdictional standards is easier.

- Using the same format as supporting agencies (e.g. State or Federal) provides compatibility among plans and may eliminate confusion in a crisis.
- Similarity creates a psychological sense of cohesiveness—multiple plans appear part of the same program.

The point is that, despite pressures to use a format of one type or another, there may be no requirement to do so. Some states have mandated the use of formats to ensure consistency and compatibility, but even most of these allow some latitude for local governments to develop plans appropriate for their communities. Consequently, planners should take advantage of this to make sure that plans are realistic and appropriate in terms of community values and available resources. If a plan doesn't work for the jurisdiction, it is unlikely to be used.

Determining Plan Content

The previous discussion suggests that, in developing an emergency operations plan, one should first consider the environment in which the plan is to function and what elements are appropriate for inclusion in the jurisdiction's plans. SLG 101 identifies five factors that should be considered during the planning process:

1. Organization—plans that are written to fulfill grant requirements are bulky and almost useless in a crisis. A plan designed for use is simple and organized so that critical information can be located quickly and intuitively.
2. Progression—plan elements should flow logically.
3. Consistency—plan elements should be formatted in the same way for ease of use and various components should not disagree or conflict with each other. This is usually a problem between the base plan and supporting annexes and plans.
4. Adaptability—as has been noted, the greater the level of crisis, the more there is a need for innovation and creative thinking. Plans should serve as a toolbox of potential solutions and be capable of adapting to unanticipated problems.
5. Compatibility—plans should enhance coordination and exclusivity. This suggests that they should be coordinated with supporting

agencies and plan stakeholders to make sure that there is congruence among various agency plans.

With these success factors in mind, one should consider how the plan should be organized. Plans usually have three basic components:

1. Base Plan—this element articulates the scope and overall concept of the plan. It also includes relevant policy and authorities and assigns responsibilities for carrying out specific actions during the crisis. It describes functional roles and responsibilities for various responding organizations and establishes lines of authority.
2. Functional Annexes—these provide detailed procedures that must be implemented to accomplish the assigned function. These annexes should include both tactical functions (e.g. building inspection) and supporting functions such as financial or logistical management. Functional annexes may also be supplemented by departmental plans, standard operating procedures, and other supporting documents.
3. Incident or Hazard Specific Annexes—these are essentially contingency plans for particular events and provide supplemental information to the functional annexes. They may actually be separate plans that are incorporated by reference.

As was pointed out in the last chapter, there are elements that a jurisdiction must consider in developing a functional plan. How these elements are addressed in the plan and the degree of detail are the decisions of the planning group and should be based on the risk assessment and a realistic appraisal of resources. These decisions should also take into account the operating level of the plan—tactical or operational—and the availability of supporting plans and documents.

USE GRAPHIC TOOLS

It is an old adage that a picture is worth a thousand words. There is, in fact, a psychological basis for this. While many people are able to read and comprehend a written description of a process, many others are visually oriented and are more likely to understand a graphical display of the same information. This is part of the reason a situation map, or status chart, proves so useful in an emergency operations center. In a time of crisis, simplicity is essential. Consideration should be given, therefore, to using simple graphical tools both in written plans and in actual operations.

One problem that frequently occurs in planning is determining whether a particular process flows logically. A good way to ensure that this occurs is to graph the processes using a flowchart. Figure 8.10 shows an example of this technique. This particular flowchart was developed to analyze the process of how reports were developed in a plans unit in an emergency operations center. The flowchart identifies the personnel involved, the key processes they perform, and critical decision points. Flowcharting in this manner has a number of advantages:

- It identifies potential conflicts or illogical sequences. Lines that cross or reverse direction generally signal a potential problem area that can be simplified.
- It can highlight and eliminate redundancy or duplication.
- It identifies dependencies both in terms of what is needed to accomplish the process and those elements that are relying on the successful completion of the process.
- The end product not only is a useful planning tool but can also be included in the plan as a graphic.

Graphical tools can also be used during operations to increase understanding of data through visual displays. Figure 8.11 is an example of a combination of tools used to display information. The chart was developed as part of a demobilization plan and is a graphical display of how the operation will be reduced over time. The upper section is a Gantt chart (a graphical representation of the duration of tasks over time) that shows how long each section will be in operation. The lower section uses a bar graph to show the reduction in staff over the same period.

A third use of graphical tools is the analysis of quantitative information. The use of graphical analysis can be a powerful tool. In his book, *Visual Explanations: Images and Quantities, Evidence and Narrative*, statistician Edward Tufte demonstrates the power of graphical analysis through two gripping cases studies pertinent to emergency managers.

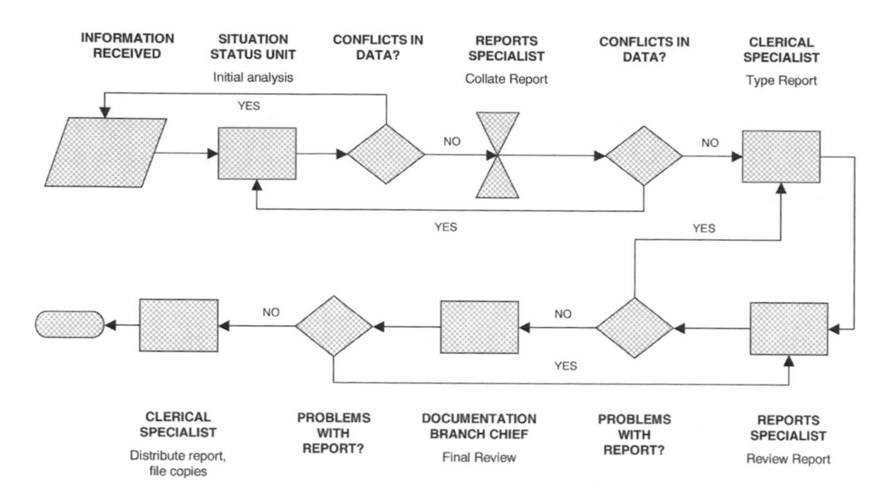

Figure 8.10 Example of flow-charting as a planning tool. This diagram traces the process for developing a situation report and identifies actors, specific actions, and key decisions. Note the logical progression and opportunity for quality control.

During a cholera epidemic in London in September 1854, Dr. John Snow was able to trace the source of the outbreak to a single pump in Broad Street by plotting known deaths on a map and conducting intensive interviews with the surviving families. Snow's work not only resulted in the end of the epidemic, it confirmed the theory that cholera was spread by contaminated water. Snow is recognized as one of the fathers of the science of epidemiology.

Tufte's second case study describes the decision making that resulted in the loss of the space shuttle Challenger. The explosion of the shuttle on January 28, 1986 was the result of two leaking O-rings that had lost their resiliency because of low temperatures. Tragically, the shuttle engineers were aware of the potential problem but were unable to convince senior officials to delay the launch. The engineers presented senior officials with 13 charts that contained sufficient data to demonstrate that launching at the expected temperatures carried the

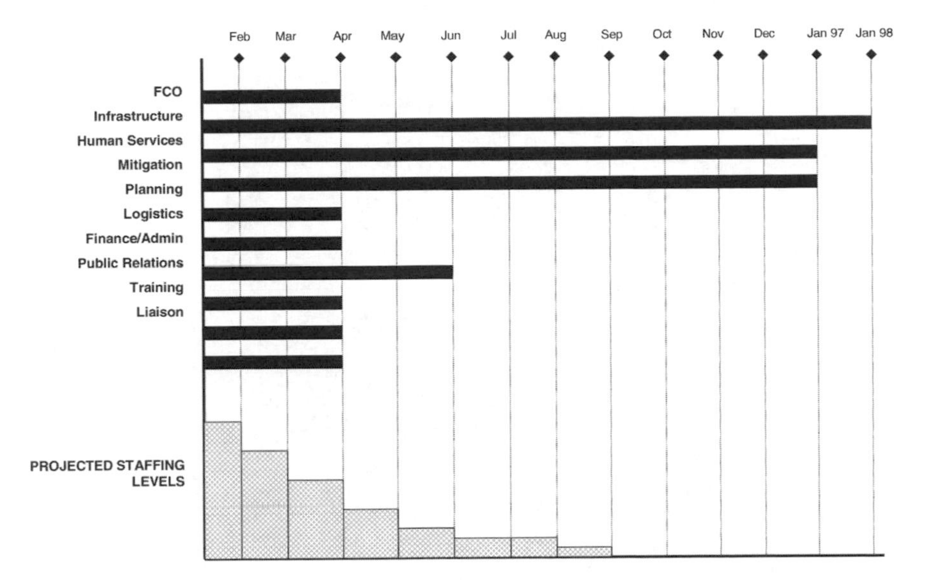

Figure 8.11 Graphic tools used to support planning. This is a graphic from a demobilization plan showing staff reductions over time. The top portion is a Gantt chart showing the length of time functions will continue to operate while the lower bar graph shows personnel reductions over time.

risk of O-ring failure. However, the data was not linked, was confusing, and, ultimately, unconvincing. Tufte demonstrates how the same data, displayed on a single chart, could have had a much more powerful impact.

Graphic tools offer the ability to provide vast amounts of information quickly and simply. Figure 8.14 shows an 1885 map graphic produced by Charles Joseph Minard that traces the destruction of Napoleon's Grand Armèe during the Russian campaign in 1812–1813. The one graphic demonstrates the effect of time, distance, temperature, and battles in a single graphic that evokes the incredible losses of the campaign. While few emergency managers can match the genius

Figure 8.12 The map produced by Dr. John Snow during the London cholera epidemic of 1854. By plotting clusters of cases based on extensive interviews with local residents, Snow traced the source of the outbreak to a single water pump.

of Minard, there is a modern equivalent—the geographic information system (GIS).

The true strength of GIS is that it can integrate disparate databases through common geospatial information and produce either list summaries or graphical outputs, most typically a map. Because GIS works

Figure 8.13 The tragic loss of the space shuttle Challenger might have been avoided if vital engineering information had been displayed in a more understandable format. *Courtesy of the National Aeronautics and Space Administration.*

in layers, it is possible to overlay data such as an electrical power grid or a sewer system on a base map to support an operational need. The system can also integrate aerial photographs or orthographic maps to provide detailed damage assessment information. For this reason, GIS products should be considered both in plan development and during operations.

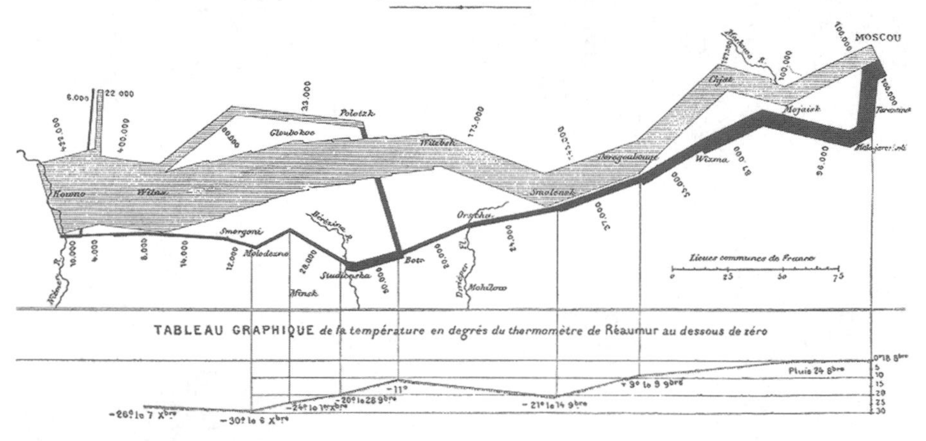

Figure 8.14 Charles Joseph Minard's 1885 map of Napoleon's Russian Campaign, probably the finest statistical graphic ever produced. The strength of the army is indicated by the thickness of the line and the bottom portion shows altitude and the temperature during the retreat from Moscow.

USE EXERCISES TO TEST CONCEPTS

Chapter 4 discussed the use of exercises as a means for providing a qualitative measurement of program capabilities. Exercises can also be used during the development of plans to test concepts before they are finalized in the plan. This testing of concepts is particularly useful if the concept under consideration is a departure from how the jurisdiction has operated in the past or has some controversial aspects to it. Discussion-based exercises, especially the tabletop exercise, are particularly useful in this regard. Tabletop exercises provide a low-stress, controlled environment to discuss plan implementation issues. Because it is a guided discussion, a skilled facilitator can maintain focus while allowing for wide-ranging exploration of the issues.

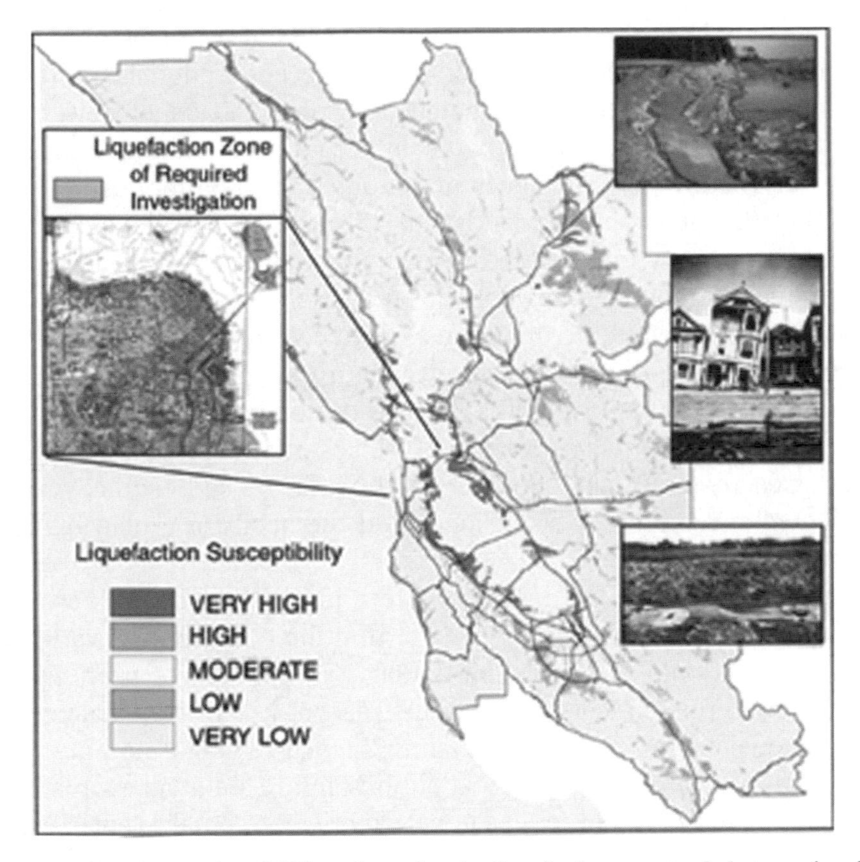

Figure 8.15 Composite of GIS products showing liquefaction zones and photographs of damages resulting from liquefaction during various earthquakes. *Courtesy of the U.S. Geological Survey.*

There are obvious advantages to testing concepts before including them in a plan. Concepts get fully discussed and issues can be identified and addressed before they become problems during plan implementation. Resource needs can be identified and included in the plan. Key stakeholders who were not considered during initial discussions can be identified. During the planning stages, people have less emotional investment in the concept and may be more willing to modify it.

An example serves to illustrate this latter point. Upon assuming her duties, a new emergency manager read the recently published jurisdictional plan and realized that it included a concept that she knew, based on her experience, could not work in actual practice. She consulted privately with other colleagues to validate her assumptions and they concurred. The problem was that the group that had developed the plan was strongly invested in the plan and was unlikely to accept the recommendations of a newcomer, no matter how experienced. The planning group was a permanent committee with which the new manager would have to work during her tenure and she was understandably sensitive to their concerns. Instead of confronting the issue head on, the emergency manager crafted a series of tabletop and functional exercises to stress that portion of the plan. The results were as expected and the group itself identified the need for a plan revision eliminating the concept.

No plan can be considered complete until it has been exercised. This is really the only way to ensure that the concepts are valid and that jurisdictional resources, particularly in the area of training, are sufficient to implement them. For this reason, it is prudent to conduct a functional exercise prior to finalizing the plan. There is nothing more embarrassing than making a great public fanfare about a new plan and then have that plan fail, either in an exercise or actual event. Testing during plan development increases confidence in the plan, validates concepts, and helps avoid future embarrassments.

KEEP IT SIMPLE

Crisis communications consultant Art Botterell has evolved four Laws of Emergency Management based on his experiences over the years at the state and federal response levels. His first and third laws are particularly relevant to planning. Botterell's First Law of Emergency Management cautions, "Stress makes you stupid." His third law reminds planners, "No matter whom you train, someone else will show up."

If one considers Boterell's two laws along with the social science literature, particularly the University of Washington study on the use of emergency plans, one fact becomes painfully obvious: it is unlikely that an emergency plan will be read at the time of crisis. This reinforces the need for aggressive exercise programs aimed at team building and developing familiarity with plans and planning concepts. However, since, as Botterell cautions, not everyone will attend training and exercises, it also suggests that plans should be simple and based on concepts that can guide creative problem solving rather than provide detailed instructions. Plans should be presented in a format that provides quick access to proposed solutions and options rather than detailed expositions of policy issues.

This hearkens back to the discussion in the previous chapter of writing plans that are appropriate for the level at which they will be used. While lengthy plans will most likely not be read, there are tools that can be extracted from the plan. Checklists have a good chance of at least being glanced at. Visual information that can be quickly assimilated such as maps and flowcharts can be useful. The best final check on a plan is to view the end product as if one is seeing it for the first time, has never been to an exercise, and has five minutes to learn their job before beginning work. One should ask, given these conditions, "What information does that person need to know and can it be quickly extracted from the plan?"

Simplicity in planning enhances the execution of the plan by focusing on results rather than process. This allows latitude for the type of creative problem solving required in disaster operations. General George Patton recommended, "Never tell people how to do things. Tell them what to do and they will surprise you with their ingenuity." This implies that the focus of planning should be to provide a solid foundation of tactical and operational tools that can then be employed as the situation dictates. Plans should provide for carrying out generic tasks but should also have the flexibility to deal with unexpected situations and emergent organizations.

CONCLUSION

Viewing planning as a process does not mean that one cannot expect results and end products. It is precisely because it is a process that it requires direction and quality controls. These controls consist of a planning structure, a decision-making process, and mutually agreeable objectives and milestones. Planning is tactically oriented. That is, it represents what the community believes is likely to happen and how it will handle the generic tasks common to all disasters. Planning can also be operationally oriented as in the case of contingency planning, providing for initial support to field operations.

Ultimately, planning is about bringing together stakeholders and sharing ideas, working out conflicts, and reinforcing strengths. It is as much about building relationships as it is about documenting policies and procedures. Consequently, planning lies at the center of the emergency management program. In considering the 2003 blackout, motivational speaker Danish Ahmed identified three important distinctions between plans and programs:

1. Plans are based on theory; programs are based on results.
2. Plans are dormant; programs are active.
3. Plans become obsolete; programs evolve.

Ahmed's insights hold particular cautions for the emergency manager. Communities that focus on plan development rather than program development are not likely to have all the supporting structures in place to make plans effective. Emergency plans are end products; emergency management programs are the process by which plans are developed, supported, and implemented.

Chapter 9

COORDINATING DISASTER

A quick and overwhelming response is better than a well planned and thought out response.
—Craig Fugate
Director, Florida Division of Emergency Management

The ultimate test of any strategy or plan is an actual event. Previous chapters have considered what social science studies suggest really occurs in disaster rather than what is commonly believed to happen and have considered methods for establishing the emergency management program and developing strategies and plans for mitigation, response, continuity, and recovery. This chapter looks at some of the issues associated with coordinating tactical response.

In coordinating tactical response, as opposed to operational response, it is well to keep in mind that disasters affect communities, not just government agencies. Ultimately, the response must involve the entire community and coordination must address not only the immediate operational concerns such as rescue, sheltering, and medical services, but must consider immediate continuity problems and long-range issues related to mitigation and recovery. Response, in its larger sense, is not just about safeguarding lives, property, and the environment, but is ultimately about restoring community viability.

TACTICAL RESPONSE

As has been discussed in previous chapters, there are several layers of community response that will be occurring simultaneously. There is the immediate operational response that happens automatically and deals with the immediate needs of victims, primarily search and rescue and emergency medical operations. There are also strategic issues that must be addressed by senior government officials. Between the two lies tactical response, the coordination of the organization and resources responding to disaster.

Where operational response considers immediate needs, tactical response must take a wider view and consider both what is occurring, but more importantly, what is likely to occur. The job of the disaster management team is to anticipate requirements in support of the operational response and to deal with the qualitative effects of disaster: convergent organizations, loss of autonomy, different performance standards, closer public private interface, and impacts on the organization. This means that the disaster management team must insulate itself from becoming overly involved in the field operation. Tactical response has two primary functions: the collection and analysis of information to assist decision making (situational awareness) and the management of resources in support of the field operation. All other activities of the disaster management team relate back to these two functions.

Operational response, to a certain extent, is predetermined. For example, a firefighter knows that he or she will be required to fight fires, conduct search and rescue operations, and provide emergency medical services. At the tactical level, the disaster management team is faced with non-routine issues and must develop solutions in limited time under crisis conditions.

Tactical response is characterized by emphasis on coordination and by creative problem solving. Craig Fugate, Director of Florida's Division of Emergency Management during the 2004 hurricanes, emphasizes this. Fugate recommends that disaster management teams be aggressive in meeting the needs of disaster victims. In his words, teams should "use a sledgehammer" as it rarely pays to be subtle. Fugate maintains that it is better to have too many resources than not enough and that resources should be pushed to affected areas without

Figure 9.1 Hurricane Charley makes landfall in the Florida panhandle August 13, 2004. This Category 4 hurricane was part of series of hurricanes that battered Florida in the 2004–2005 hurricane season. Florida's response to the storms is an example of emergency management at its best. *Courtesy of the National Oceanic and Atmospheric Association.*

waiting for requests. The longer one waits to gather facts, the harder it is to change the outcome, according to Fugate.

Chapter 7 considered 10 principles for disaster planning developed by Dr. E. L. Quarantelli. One of those principles is that there is a difference between disaster preparedness planning and disaster management. Quarantelli has also developed ten principles for disaster management that serve to guide tactical response:

1. Recognize the difference between agent and response generated needs—there are needs that are driven by the type of hazard that

created the disaster and others that are generated by the response itself. The former is dependent on the impact and nature of the hazard, such as the requirement for mass prophylaxis during a pandemic or biological attack. The latter are fairly constant from disaster to disaster. For example, there is always a need to coordinate the activities of emergent groups or to order and transport critical resources. Agent-generated needs will normally require a situational approach; they can be pre-planned to a certain degree but actual response will be situation dependent. Response-generated needs, on the other hand, can be pre-planned as part of the community's emergency management program.

2. Carry out generic functions in an adequate way—this is the essence of the functional approach to emergency planning discussed in Chapter 7. It is possible to identify response-generated needs for specific functions such as evacuation or sheltering and to pre-plan generic responses for these functions. Figure 9.2 is a list of operational functions from NFPA 1600. This list is repre-

- Control of access to the area affected by the disaster
- Identification of personnel engaged in activities at the incident
- Accounting for personnel engaged in incident activities
- Accounting for persons affected, displaced, or injured by the disaster/emergency
- Mobilization and demobilization of resources
- Provision of temporary, short-term, or long-term housing, feeding, and care of populations displaced by a disaster/emergency
- Recovery, identification, and safeguarding of human remains
- Provision for the mental health and physical well being of individuals affected by the disaster/emergency
- Provision for managing critical incident stress for responders

Figure 9.2 List of operational procedures from NFPA 1600, Annex A. Pre-planning can be done for these types of response-generated needs.

sentative of the types of operational-level activities that lend themselves to pre-planning.

3. Mobilize personnel and resources in an effective manner—as has been noted, a characteristic of disaster is the overwhelming amount of resources that are mobilized and made available to the affected community. The challenge is the effective deployment of those resources to where the need is greatest. An important part of disaster response is matching resources against need as quickly as possible. For this reason, resource management and logistics are essential, but often neglected, components of emergency plans.
4. Involve proper task delegation and division of labor—an important part of coordination is the ability to identify and delegate tasks. Many tasks can be pre-identified and the responsibility assigned to an agency or department. However, at the time of a disaster, many new tasks emerge and they there must be a mechanism to delegate these tasks to specific action agents.
5. Allow the adequate processing of information—many organizations are not equipped to expand information collection, analysis, and dissemination beyond the systems used in normal day-to-day operations. There is a considerable amount of information both required and available after an event as more and more agencies become involved. Unfortunately, much of the information may be conflicting and there may be gaps in the data. Failure to manage information flows will degrade what is known as "situational awareness," an understanding of the overall nature of the event, and lead to misapplication of resources or failures in response.
6. Permit the proper exercise of decision-making—disasters create requirements for new, non-traditional tasks and conflicts between existing and emergent groups. This can lead to conflicts over who has the authority to make decisions or commit resources. An important part of tactical response is facilitating the decision-making process. This process is a combination of good information about the situation and available resources and an environment that fosters cooperative decision making.

7. Focus on the development of overall organizational coordination—as has been mentioned, the goal of tactical response is not control but coordination. This is done by establishing an inclusive operating structure and facilitating identification of common objectives. While agencies will operate independently and under their own authorities, the development of a coordinated action plan assures that these decentralized operations achieve tactical objectives.
8. Blend emergent aspects with established ones—as was noted in Chapter 2, communities are resilient and new organizations will arise in the wake of a disaster. The disaster management organization must recognize these new organizations and incorporate their activities into the response. The tendency to ignore emergent groups, particularly those from the private sector, can eliminate potential resources and generate ill will as the operation moves towards recovery.
9. Provide the mass communication system with appropriate information—the principle mechanism for providing disaster information to the public is through the media. However, if the information provided is inaccurate or inadequate, the credibility of the response organization will be diminished. An important function of the disaster management team is making sure the media is integrated into the response in an appropriate way.
10. Have a well functioning Emergency Operations Center—the EOC is the central coordination point for the response effort. If that point fails, centralized coordination dissolves into chaos.

Quarantelli's 10 principles reinforce the point made in previous chapters: disasters are not just about life safety response; they are multi-agency/multi-jurisdictional events that must be managed through coordination, not command and control. The principles are not, per se, functions that must be included in an emergency plan, but are instead common threads that run through all plans. They underscore the need to develop functional plans with clear assignment of responsibilities. They reinforce the need for pre-planning. Most importantly, they stress the need to create mechanisms that allow for coordinated decision making and cooperative planning.

INCIDENT MANAGEMENT SYSTEMS

Small, specialized teams may be fully capable of dealing with day-to-day emergencies. However, as more resources are committed to a crisis, there is a need to impose some sort of operating structure to keep the response efficient. As has been noted, disasters are characterized by an influx of agencies and organizations that may not work together on a regular basis, increasing the risk of confusion and inefficient application of resources. A system for coordinating and managing resources is essential, and, if no such system exists, one will eventually evolve over time.

For public sector agencies, the National Incident Management System (NIMS), developed in 2004, mandated the use of the Incident Command System (ICS). ICS originated in Southern California in response to a series of wildfires in 1970 that burned over 600,000 acres and claimed 16 lives in only 13 days. The fires affected jurisdictions at the federal, state, and local levels simultaneously and highlighted the barriers preventing multiple responding agencies from working together. In 1971, Congress provided funding for the United State Forest Service to establish a working group to analyze the problems from the fire and to create a working solution. The working group, which would eventually become known as FIRESCOPE (Firefighting Resources of Southern California Organized for Potential Emergencies), identified six major problem areas:

- Lack of a common organization—the multiple agencies responding used different terminology and organizational structures, creating confusion.
- Poor on-scene and interagency communications—radio systems and operating protocols were different, making information sharing among agencies almost impossible.
- Inadequate joint planning—agencies operated from autonomous command posts, with no joint planning for operations or logistics.
- Lack of valid and timely intelligence—no one was assigned responsibility for intelligence collection and collation and decisions were made on the basis of incomplete, outdated, or inaccurate information.

- Inadequate resource management—vital equipment was lost or misplaced, there was no common system for ordering resources, and operations were over- or under-staffed.
- Limited prediction capability—with the lack of intelligence, confusion in the deployment of resources, and non-existent joint planning, it was impossible to predict future requirements such as the requirement for evacuation or the need for additional resources.

It was clear from the analysis of the fires that there was a need for an incident management system that could integrate the activities of multiple responding agencies. The working group next developed design criteria for the system. The working group determined that the system must:

- Be able to handle incidents involving single agencies up to those involving multiple agencies and jurisdictions
- Be able to expand from use in daily incidents to major emergencies
- Be usable in all emergencies, not just fires
- Be able to integrate new technologies as they are developed
- Have standardized terminology, organization, and procedures
- Integrate with existing agency procedures with minimal disruption
- Be simple to teach and maintain.

It is important to understand the reasons behind ICS and what FIRESCOPE was trying to develop before one gets involved in the nuts and bolts of ICS. Too often, planners tend to focus on the operational structure developed under ICS and assume that, because the structure is in place, ICS is being used. The intent of the FIRESCOPE planners was not to develop an operating structure but an operating system. Most mis-applications or failures of ICS can be traced to a lack of understanding of the underlying concepts and principles of the system.

ICS is based on four important concepts: agency autonomy, unit integrity, functional clarity, and management by objectives. The FIRESCOPE planners were drawn from a wide range of federal, state, and local agencies and they understood the barriers that could create problems at an incident. Principal among these was the issue of jurisdictional autonomy and statutory responsibility. Consequently, the system is designed in such a way that organizations with overlapping jurisdictions can still meet their legal and fiscal obligations while working as part of an ICS team.

Similarly, the system is intended to ensure that personnel from the same agencies and disciplines continue to work together. It is a common misconception that using ICS requires accepting orders from someone with little or no knowledge (e.g. a police officer directing a firefighter how to fight a fire) or that responding units will be broken up before being deployed. Unit integrity is maintained and most responders receive orders through their existing chain of command.

Functional clarity is the result of the functional approach to crisis discussed in Chapter 7. An individual or unit need only focus on their assigned function while other members of the organization handle supporting tasks. Because ICS uses common terminology, functional assignments are standardized and easily understood.

Management by objectives is a business concept adapted for use in emergency management. First espoused by Peter Drucker in 1954 in his book, *The Practice of Management*, management by objectives is the development of specific, measurable objectives that are agreed to by members of the organization and then monitored to ensure they are achieved. This is the same system that was used in Chapter 4 in developing the strategic plan and was applied to the development of work plans in Chapter 8.

In addition to management by objectives, ICS borrows other business concepts, such as manageable span of control, and concepts from the military (e.g. unity of command). Figure 9.3 lists the management characteristics of ICS. These characteristics comprise the heart of ICS and are frequently overlooked. Taken as a whole, they form a system that establishes a robust operational framework for dealing with crisis. Further, these principles are effective even where the field command and control structure is not used, such as at the EOC.

Figure 9.4 shows the ICS field command structure developed under FIRESCOPE. Over the years this model has been adapted and fine-tuned by different organizations to meet their unique needs. For example, Figure 9.5 shows an adaptation of the ICS structure used in hospitals. There are also specific adaptations for law enforcement and for private industry. Figure 9.6 shows the ICS structure adapted for use by

Common terminology

Modular organization

- •Establishment and transfer of command
- •Chain of command and unity of command
- •Unified command
- •Unity of command

Management by objectives

Reliance on an Incident Action Plan

Manageable span of control

Comprehensive resource management

- •Pre-designated incident locations and facilities
- •Deployment
- •Accountability

Integrated communications

Information and intelligence management

Figure 9.3 Management characteristics of the Incident Command System. This is a slightly modified rendering of the list in the National Incident Management System.

a large bank. This infinite adaptability of the ICS structure suggests several points. The first is that, contrary to the arguments that have been waged over the years, there is no "pure ICS" structure. The original FIRESCOPE model was designed specifically for large wildland fires involving multiple jurisdictions, but it was always intended to be modular and adaptable to changing circumstances. This is reinforced in NIMS, which specifically recognizes the right of the Incident Commander "…to modify procedures or *organizational structure* to align as necessary with the operating characteristics of their specific jurisdictions or to accomplish the mission in the context of a particular hazard scenario." So long as the five key functions of command, plan-

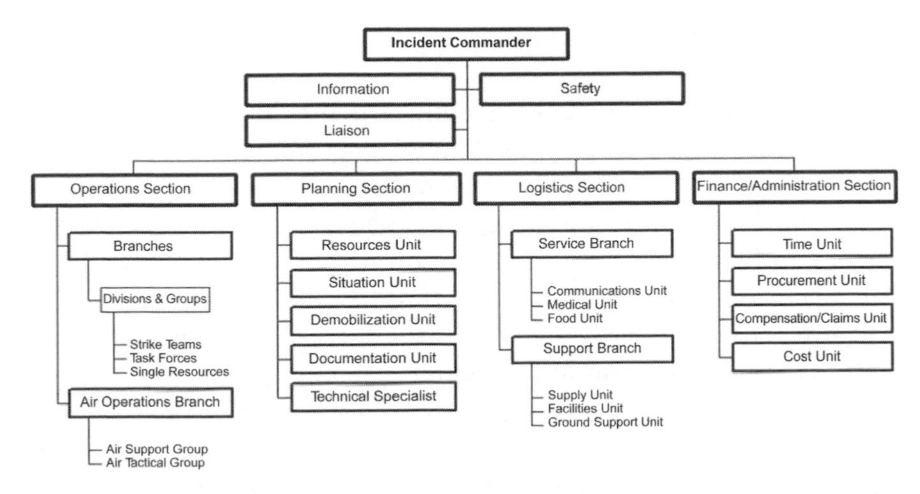

Figure 9.4 The standard ICS structure as developed under FIRESCOPE. Courtesy of Regina Phelps, EMS Solutions, Inc.

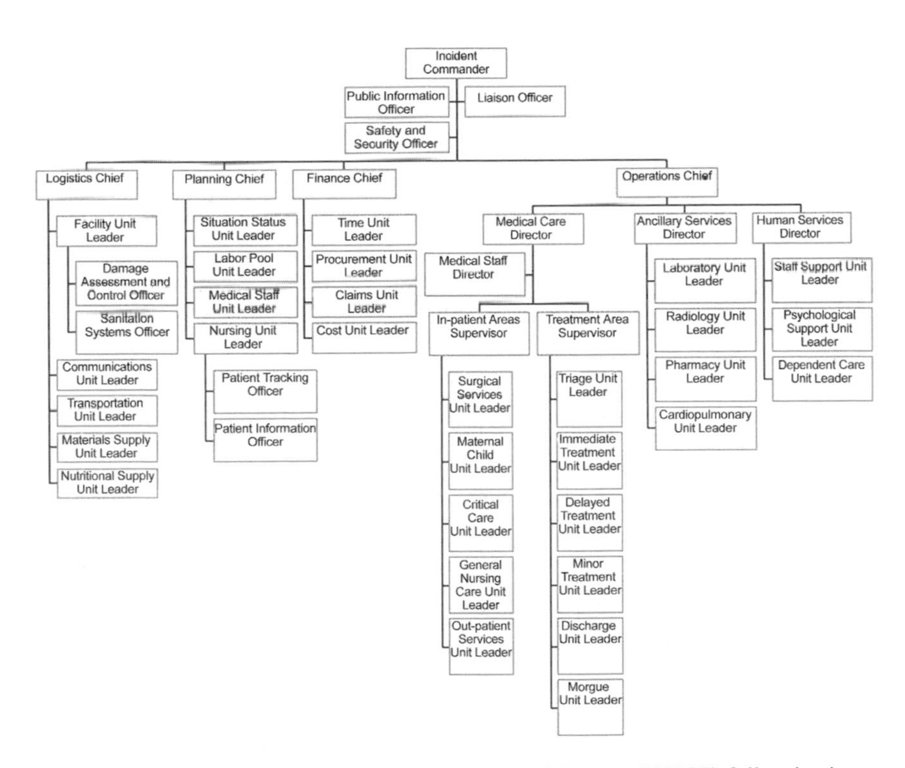

Figure 9.5 Hospital Emergency Incident Command System (HEICS) full activation.

ning, logistics, finance, and operations are addressed, the structure can be adapted as needed.

The adaptability of the ICS operating structure also suggests that, to a certain extent, it is irrelevant to the successful implementation of ICS. Under ICS, the incident commander is responsible for performing all functions that are not delegated. The modularity of the structure comes into being when the incident commander begins to apply the principle of span of control by delegating functions. Similarly, success is determined not by having all boxes on the organization chart filled, but by determining common objectives, managing resources effectively, and providing clear direction. It is a cardinal mistake to assume that using the ICS operating structure is the equivalent of using the Incident Command System just as it is a mistake to believe that because one is using ICS that one is using the National Incident Management System.

Like ICS, NIMS has been subject to a lot of misunderstanding. Part of this is the result of the mistaken belief that one need only adopt the

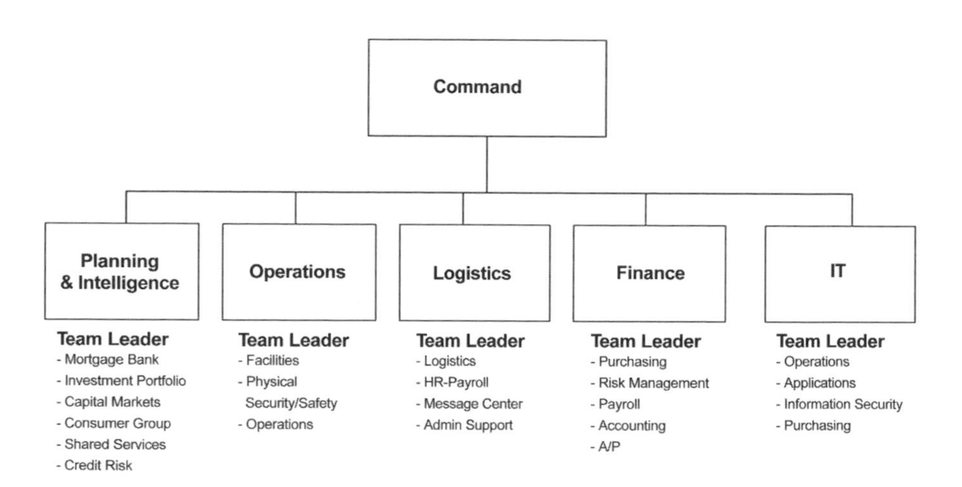

Figure 9.6 The ICS operating structure adapted for use by a large banking company. Courtesy of Regina Phelps, EMS Solutions, Inc.

ICS operating structure to be compliant. Like ICS, NIMS is a system that was established to support response operations. ICS, although it receives the bulk of the treatment in the NIMS document (80+ pages of 139 pages), is only a small element of a much larger system that includes many components that are integral to the emergency management program such as planning, training, and exercising. Compliance with NIMS, therefore, is generated not through the adoption of the ICS operating structure but through an effective emergency management program.

Unified and Area Commands

As has been noted, one of the concepts behind the Incident Command System is agency autonomy—the recognition that agencies and jurisdictions have statutory responsibilities that cannot be surrendered to a crisis management organization. To deal with circumstances where jurisdictional authorities overlap, ICS incorporates a concept called unified command.

The term "unified command" has become extremely common, especially in the wake of Hurricane Katrina, but it is one of the most misunderstood and misused terms in ICS. Many officials use unified command to mean a group of senior officials or agency representatives gathered in a single operations center. However, unified command is a process that allows all agencies with responsibility for an incident, either geographical or functional, to establish a common set of incident objectives and strategies without abdicating agency authority, responsibility, or accountability.

Critical to the formation of a unified command is recognition of who has jurisdictional authority for the incident. The incident may affect multiple jurisdictions, involve multiple agencies from the same jurisdiction, or have an impact on multiple jurisdictions or agencies. The key here is jurisdictional responsibility—just because an agency provides resources to assist in responding to the incident does not automatically make them part of a unified command. Further, while there may be multiple incident commanders, a unified command consists of:

- A single, integrated incident organization
- Collocated (shared) facilities

- A single planning process and incident action plan
- Shared planning, logistical, and finance/administration operations
- Coordinated process for resource ordering

In other words, a unified command may have multiple commanders but will act and respond as a single organization using ICS principles. The two concepts of limiting leadership to only those with jurisdiction and creating a single organization are the two most common failures under unified command.

Where unified command describes the command relationship among agencies responding to an incident, an area command is established to oversee the management of multiple incidents that are each being handled by a separate ICS organization or a very large incident that involves multiple ICS organizations. An area command oversees management of the incident or incidents under its purview by:

- Setting overall incident-related priorities.
- Allocating critical resources based on priorities.
- Ensuring that incidents are properly managed.
- Ensuring that incident(s) objectives are met and do not conflict with each other or with agency policy.

It is important to note that an area command retains a direct command and control relationship with the other ICS organizations. The focus is on management of the incident and the resources related to the incident. In other words, an area command is still operational in nature in that it focuses on the successful resolution of the incident or incidents, not on long-term impacts or potential needs.

The decision to establish an area command is normally based on the ICS principle of manageable span of control and on the geographical dispersion of the incident or incidents. This decision is separate from the decision as to the command structure—an area command can be headed by a single area commander or by a unified area command.

Multi-agency Coordination Systems

ICS has, to a certain extent, become a victim of its own success and is considered the solution to managing disasters at all levels. Thus, ICS is the basis of both the National Incident Management System (NIMS) and the National Response Plan and there is a national effort underway to force its adoption across the United States. There is no question that, as an operational tool, ICS is extremely effective. However, there is no systematic empirical research on the effectiveness of ICS as a mechanism for coordinating disaster response and the failure of the National Response Plan in Hurricane Katrina raises questions about its use.

ICS has worked well for two main reasons. First, most responders operating under the system are from paramilitary organizations and are used to a hierarchical command and control system. Secondly, operations that have seen successful use of ICS are predominantly single events or multiple events of the same type and required a response that was essentially operational in nature. For example, the Pentagon attack on September 11 was successfully managed under ICS, as was the Oklahoma City bombing, once training was provided to local agencies. There are also examples of ICS overhead teams being successfully used to assist local governments in organizing operational response, most recently in Hurricane Katrina.

As has been noted, disasters differ from emergencies both quantitatively and qualitatively. As the size of the event grows, more and more agencies are drawn into the response and issues of continuity and recovery begin to compete for resources. At the same time, disruption of communications systems and normal operating structures make the type of hierarchical approach use by the Incident Command System difficult to maintain. There is a need for the coordination of decentralized and possibly even independent actions rather than for direct operational control of the incident. Consequently, at the tactical response level, one uses a component of ICS called the multi-agency coordination system (MACS) rather than the field level organizational structure shown in Figure 9.4.

This point is one that frequently confuses planners, including those involved in developing the National Response Plan. One hears, for example, Michael Brown, former director of FEMA, responding to questions on the perceived failure of his agency in Hurricane Katrina, saying, "I was unable to form an area command." This statement demonstrates a major flaw in the National Response Plan and a lack of a basic understanding of operational dynamics. Area command, as was

CASE STUDY: JACKSON COUNTY, MISSISSIPPI—USING ICS PRINCIPLES TO COORDINATE DISASTER

Jackson County is a small county located 68 miles east of New Orleans. Consisting of 727 square miles and with a population of 133,000, the county is largely rural, with most of the population concentrated in coastal communities. Hurricane Katrina damaged or destroyed over 29,000 residences and caused considerable infrastructure damage. With few resources to manage the disaster, the county requested help from the Mississippi Emergency Management Agency, who in turn asked for help through the Emergency Management Assistance Compact.

In response to the EMAC request, California dispatched the East Bay Incident Management Team. An incident management team is a group of trained responders specializing in the incident command system with defined tasks or functional responsibilities. Teams are trained to address the key functions of operations, planning, logistics, and finance in the context of all risks.

(continued on next page)

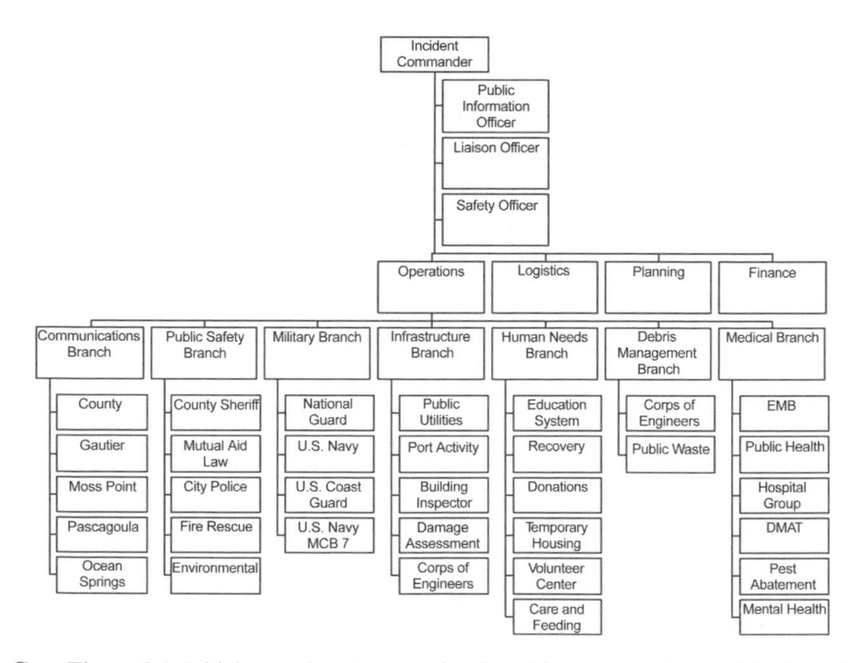

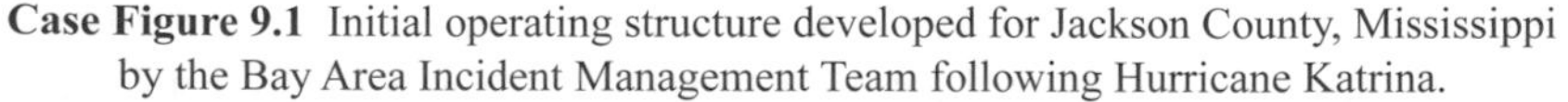
Case Figure 9.1 Initial operating structure developed for Jackson County, Mississippi by the Bay Area Incident Management Team following Hurricane Katrina.

(continued from previous page)

When the full team assembled in Jackson County on September 15, they were faced with chaos. Public and private agencies were operating independently, there were control issues between the cities and the county, multiple points existed for ordering resources, and public safety infrastructure was severely damaged. Worse, many local government employees were themselves victims, and while some incident objectives had been identified, coordination was not being done with the agencies tasked to accomplish them. It was obvious that the county was still responding to the disaster.

The team recognized that it would first need to transition the current system from independent action to facilitated decision making and move from response operations to recovery. Once this was accomplished, the goal would then be to gradually transition operations to local control. With this end in mind, the team began conducting facilitated strategy meetings focused on unified decision making and establishing country-wide policy. Eventually, the team evolved an organizational structure that facilitated interagency coordination.

(continued on next page)

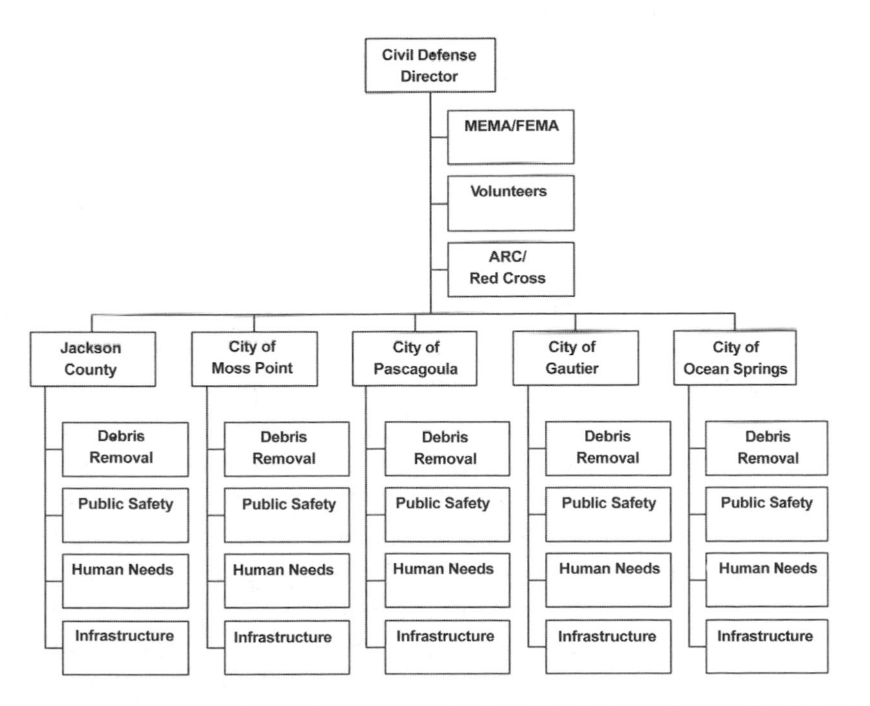

Case Figure 9.2 Final operating structure developed for Jackson County, Mississippi by the Bay Area Incident Management Team following Hurricane Katrina. Note the shift from an organizational function based on function to one based on geography.

(continued from previous page)

As part of its effort, the team also focused on creating a single point for ordering and tracking resources. One of the first problems the team identified was a bottleneck that prevented resource requests from being processed because of a single, overwhelmed official. As operations became smoother, the team integrated local players into the ICS organizations and trained them on functional tasks. Eventually, the team evolved a second structure to complete the transition to local control for recovery.

The success of the East Bay Incident Management Team in Jackson County was based on a number of factors. First, the team was cohesive, experienced, and well trained. They approached the problem with an agreed upon methodology and a clearly defined end-state of transitioning operations to local control. However, rather than trying to impose a rigid organization on the county, the team used ICS principles to guide the evolution of an organizational structure that worked for the stakeholders. The organizational structure was effective because it was developed by the stakeholders and took into consideration the local political environment and existing relationships and processes. It was also flexible enough to accommodate emergent organizations and outside agencies.

discussed previously, is an operational concept used to manage multiple incidents tasked to separate ICS organizations or to oversee a very large incident involving multiple ICS organizations. It is a command and control function. It is unlikely that Director Brown was giving orders to independent federal agencies or to the State of Louisiana. MACS, on the other hand, is a system for coordinating and supporting incident management activities. It does not attempt to manage field operations directly but is focused on establishing common objectives and removing barriers to successful response.

The difference between MACS and the field operating structure of ICS is more than semantic. Under ICS, responding units come under the operational control of the incident commander or the unified command. In MACS, the EOC director coordinates the activities of independent agencies, the public, and the political leadership of the jurisdiction but has no operational control over them. There is no attempt to direct the activity of independent agencies. Rather, MACS is used to make sure that the activities of those agencies are consistent with common incident objectives and that each agency is aware of what the other is doing. MACS has five main responsibilities:

- Incident priority determination—a hard decision in any disaster is where to place priority for services. This decision should be based on an assessment of need and not driven by the media reports or the demands of politicians. These are not necessarily “either-or” decisions—relief agencies need to agree on what must be accomplished in the short-term to meet the needs of the victims, and this may involve compromise and improvisation.
- Critical resource use priorities—during the initial phases of a disaster, critical resources are scarce. As the operation progresses, resources begin to arrive but may still be in short supply for some time. Decisions must be made about how these scarce resources are distributed and who has priority for arriving resources.
- Communications systems integration—responding agencies frequently operate through multiple incompatible communications systems. However, there are ways to work around some of these incompatibilities and a joint communications plan goes a long way to resolving these issues.
- Information coordination—the coordination of information to the public through a single Joint Information Center helps to build confidence in the response and diffuse potential problems.
- Intergovernmental decision coordination—this is the main function of the MACS: the development of joint objectives and a common Incident Action Plan that provides central coordination of decentralized field operations.

Research supports the idea that the military-type model espoused in the ICS field structure may not be possible in large-scale events because of the multiple actors involved. Many of these responding organizations are not paramilitary and do not have a hierarchical structure. Others operate under their own authorities and, while willing to work cooperatively, will not surrender their autonomy to an Incident

Commander. Further, use of such a system is based on unspoken assumptions that individuals and social structures are unable to cope with the event and that only those in authority can make decisions. This forces the creation of an artificial social structure and a closed system designed to overcome this breakdown of social order by imposing a mechanism for command and control. In essence, this system is based on many of the disaster myths discussed in Chapter 2.

Command and Management
- Incident Command System
- Multi-Agency Coordination System
- Public Information Systems

Preparedness
- Planning
- Training
- Exercises
- Personnel Qualification and Certification
- Equipment Acquisition and Certification
- Mutual Aid
- Publications Management

Resource Management

Communications and Information Management
- Incident Management Communications
- Information Management

Supporting Technologies

Ongoing Management and Maintenance

Figure 9.7 Components of the National Incident Management System (NIMS).

Researchers, most notably William Waugh, have raised concerns over the fact that ICS is based on management principles that are over 30 years old. More recent business practices put much less emphasis on top-down direction and command-and-control mechanisms. There are concerns that imposing the use of ICS through NIMS may limit flexibility in response and limit the development of new, more effective models. In actual practice, communities are remarkably resilient and are capable of participating in response. Further, in a disaster of any magnitude, there will be participation by agencies from all levels of government—federal, state, and local—and issues will be influenced by politics and public expectation as much as by operational need. Given the multiplicity of actors and the changing political and operational climate, tactical response must be flexible, innovative, creative, and participative.

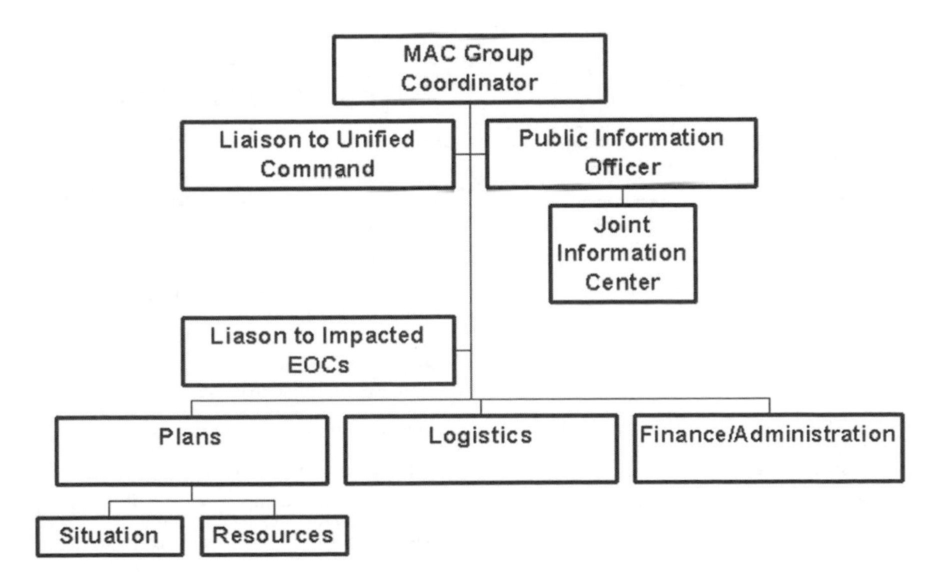

Figure 9.8 Regional Multi-Agency Coordination Group. This variation eliminates the operations section as the group does not provide command and control, only resource support.

Further confusing this issue is that MACS, being part of ICS, uses the same principles as the field components. Most crisis teams adopt the five basic functions of command, operations, planning, logistics, and finance/administration as their base operating structure, which means that the structure of a MACS looks very much like the field operating structure. Hence, there is a tendency to impose the same directive relationships that are vital in a field operation. However, at the MACS level the units and branches used at the field level are generally replaced by functional groups, such as the Emergency Support Functions found in the National Response Plan. Autonomous agencies and community groups are more likely to work together as part of a collaborative effort than under the direction of a branch chief or unit leader.

MACS is a system and, like ICS and NIMS, this is sometimes lost in the discussion. As a system, it is not tied to a particular facility or methodology. MACS can be invoked through conference calls,

MACS	Area Command
-Expansion of the off-site coordination and support system.	-Expansion of the on-site command function
-Members are agency administrators or designees from the agencies involved or heavily committed to the incident	-Members are the most highly skilled incident management personnel.
--Organization generally consists of the MAC Group (agency administrations), MAC Group Coordinator, and an intelligence and information support staff.	-Organization generally consists of an Area Commander, Area Command Planning Chief, and Area Command Logistics Chief.
-Is the agency administrator or designee.	-Is delegated authority for specific incident(s) from the agency administrator.
-Allocate and reallocate critical resources through the dispatch system by setting incident priorities.	-Assign and reassign critical resources allocated to them by MAC or the normal dispatch system organization.
-Make coordinated agency administrator level decisions on issues that affect multiple agencies.	-Ensure that incident objectives and strategies are complementary between Incident Management Teams under their supervision.

Figure 9.9 Comparison between MACS and area command. Based on training materials from the National Wildfire Coordinating Group.

occupy stand-alone facilities such as conference rooms or offices, or be established at an emergency operations center. It may consist of a single discipline such as the wildfire coordination groups used in California. One can make the argument, for example, that the law enforcement Joint Operations Center (JOC) called for in the National Response Plan is a MAC group.

The use of management by objectives as reflected in the Incident Action Plan is a critical component of coordination in the MACS. As noted in the case study on Jackson County, teams trained in ICS have adapted the system to help coordinate response and recovery activities at the jurisdictional level. However, there are critical differences between the ICS field components and MACS:

- MACS is not a command function; it is a coordination function.
- MACS does not focus on operational response; it focuses on supporting operational response by anticipating and prioritizing requirements.
- MACS is primarily oriented toward the development of strategies rather than on the development of short-term action plans.
- MACS operations generally do not provide direct services but are primarily logistical in nature. That is, MACS locates and provides the supplies and specialized teams used by the incident commander; it does not directly control those resources once they have passed to the incident commander.

Recognizing that there is a difference between controlling response (i.e. the direct provision of relief services) and the coordination of response acknowledges the qualitative differences between day-to-day response and disasters. Controlling response is operational in nature and very focused on immediate and short-term needs. Coordinating response requires looking at the big picture, anticipating future requirements, and delegating responsibility.

This suggests that what is occurring in most EOCs may, in fact, add to confusion at the time of crisis. Most jurisdictions are organized along the lines of Figure 9.10 with an EOC staff of agency representatives receiving guidance from a policy group of key department heads. However, these same department heads are direct reports to the senior

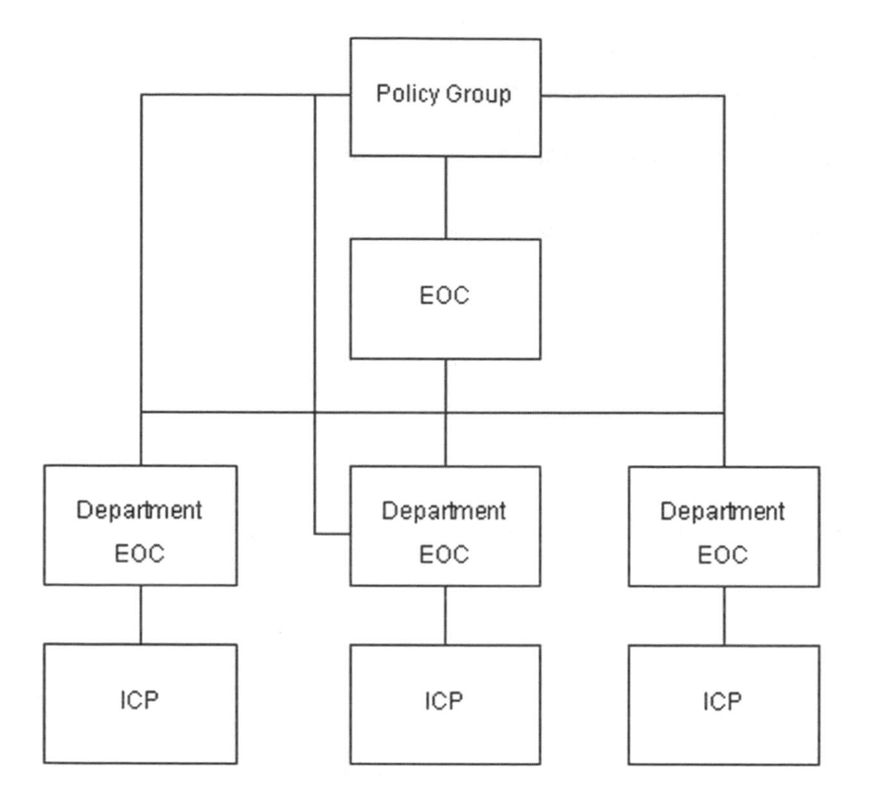

Figure 9.10 Typical community response structure. Note the conflict in command and control between the EOC and the senior officials in the policy group.

elected official and will either end up operating directly from the EOC as an impromptu unified command or will bypass the EOC because of their direct relationship. One can certainly make an argument for unified command at the EOC in the initial phases of the operation. However, operational responders tend to neglect other critical response elements related to continuity and recovery in favor of an overwhelming commitment to life safety. Further, there is usually little provision for integrating emergent or convergent organizations into this traditional structure. Ultimately, departments answer to depart-

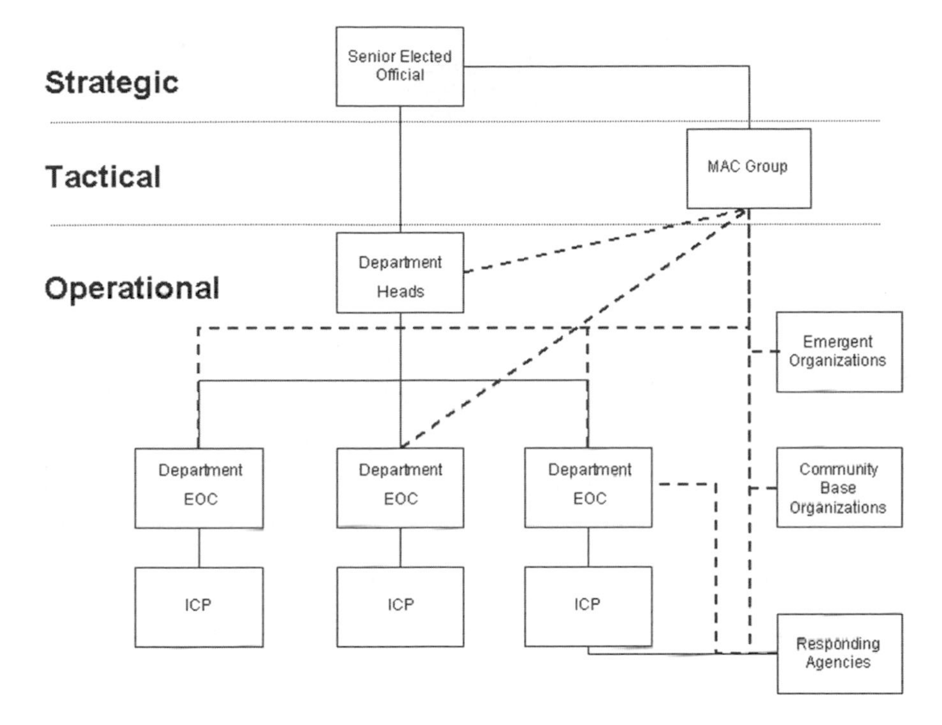

Figure 9.11 Community response structure distinguishing between command and control relationships and coordination under MACS.

ment heads who, in turn, answer to the chief elected official. The traditional EOC may actually add a layer of complexity during crisis.

Using MACS can open this closed system to a certain extent by distinguishing between coordination and command functions. Reporting relationships remain the same and command and control is accomplished through the same mechanisms that drive day-to-day operations. The MACS as a coordinating body does not threaten these relationships and is not an artificial construct that heightens stress. The closer a response structure resembles the day-to-day operational structure, the more likely people are to respond effectively to crisis.

Incident Command	Unified Command	Area Command (Unified Area Command)	Multiagency Coordination System (MACS)	Emergency Operations Center (EOC)
•Management system used to direct all operations at the incident scene. •The Incident Commander (IC) is located at an Incident Command Post (ICP) at the incident scene.	•Application of ICS used when more than one agency has incident jurisdiction. • Designated agency Incident Commanders work from a single ICP •Uses a common set of objectives and strategies •Uses a single Incident Action Plan	•Established as necessary to provide **command authority** and coordination for two or more incidents in close proximity. •Works directly with Incident Commanders •becomes Unified Area Command when incidents are multi-jurisdictional. •May be established at an EOC or at some location other than an ICP.	•System used to **coordinate resources and support** between agencies or jurisdictions. •A MAC Group functions within the MACS •Interacts with agencies or jurisdictions, not with Incident Commanders •Useful for regional situations. •Can be established at a jurisdictional EOC or at a separate facility.	•The EOC is a facility used in varying ways at all levels of government and within private industry to provide coordination, direction and control during emergencies. •EOCs may house Area Command and MACS activities as determined by agency or jurisdiction policy

Figure 9.12 Chart summarizing the command and coordination structures under the Incident Command System. Note the distinction between direct command authority and coordination of supporting resources. Based on training materials from the National Wildlife Coordinating Group.

Emergency Operations Centers

It is important to distinguish between the emergency operations center (EOC) itself, which is a facility, and the management group that occupies it. Depending on the designation of this group, the nature of the EOC changes. An EOC occupied by an area command is part of the command function and deals directly with incident commanders. The same EOC occupied by a MAC Group coordinates the activities of agencies and jurisdictions; it does not direct the activities of incident commanders or unified commands. As was noted above, for most jurisdictions, the crisis team at the EOC coordinates multi-agency and multi-jurisdictional activities and does not provide direct command, making MACS particularly suited for use in the jurisdictional EOC.

Crisis communications consultant Art Botterell has defined the emergency operations center as the place "where uncomfortable officials meet in unfamiliar surroundings to play unaccustomed roles making unpopular decisions based on inadequate information in much too little time." While a bit tongue in cheek, Botterell's description points out some of the problems common to many government officials. They are usually uncomfortable and unfamiliar with the EOC and their roles because they have never attended an exercise. Lack of information and time limits usually are the result of a failure of the EOC staff to develop an adequate plans unit, a failing that seems to have created problems for all levels of response in Hurricane Katrina.

The Emergency Management and Response-Information Sharing and Analysis Center (EMR-ISAC) sponsored by the Department of Homeland Security suggests four critical design criteria for EOCs:

- Location—for many years, the State of California's EOC was located in a flood plain, an obvious draw back to an organization that had to respond to annual flooding. One also questions the decision of the city of New York prior to September 11 to place its EOC in a complex that had been the target of a previous terrorist attack. Both of these decisions may have been made for good reasons, some of them political and some of them fiscal. However, in retrospect, locating a community's principal response facility in a location that puts it at risk from the very crisis it was built to counter seems questionable at best.
- Redundancy—NFPA 1600 and the EMAP standard require the identification of alternate operating facilities. Redundancy also applies to critical systems, such as communications and electrical power. One major city's EOC has two generators, the capacity to install a third in the future, a hook-up for attaching a portable generator to the facility, and two separate commercial feeder lines. (Ironically, this same EOC lost all power one day due to a single non-interruptible power supply switch that was inadvertently turned off when a box fell on it. This embarrassment resulted in the switch being covered and the system redesigned to remove the single point of failure.)
- Self-sufficiency—there is every reason to believe that an EOC will be cut off from outside assistance for several days. As was

noted in a previous chapter, the EOC for the city of New Orleans ceased to function when it ran out of fuel for its generator. Fuel for generators was also an issue in many Bay Area cities during the Loma Prieta earthquake in 1989.

- Communications—without the ability to gather information and disseminate instructions, the EOC cannot function. Reliance on any one communications method, such as the standard telephone system or computer-based radios, creates a vulnerability that could prove fatal to the EOC's mission. Parallel methods of communication are essential. One jurisdiction has even developed plans for using bicycle messenger services to supplement technology.

In addition to the four design criteria suggested by EMR-ISAC, Art Botterell suggests adoption of a common design language for EOCs and has identified five basic design types:

1. Boardroom—this layout consists of a single table or a horseshoe assembly of tables where participants sit facing each other. Displays are normally positioned at one end. This layout emphasizes communication and coordination and can be effective for small operations and policy groups. However, the design makes it difficult to continue work during briefings and usually does not provide enough working space for participants.
2. Mission Control—this design is technology-oriented and uses large visual displays. Participants are usually seated in rows facing the displays and interactions are through a technological "knowledge-base." This layout is effective for technical tasks but tends to diffuse interaction among the participants. Face-to-face interaction usually takes place outside the main room rather than within it.
3. Marketplace—this layout consists of multiple stations organized around supporting functions. Specialists are grouped together by function and coordination is accomplished by communicating among the functional groups.
4. Bull's Eye—this layout creates a center table or tables with supporting staff occupying concentric circles around the central

tables. This model emphasizes the importance of the key players as representatives of large agencies and facilitates staff consultation. However, it requires a considerable amount of floor space and usually is not found in local EOCs.

5. Virtual—in this day and age, Internet connectivity is critical to an EOC. With Internet connectivity, the EOC has access to information and communications. The Internet also affords the ability to conduct operations without all the players being physically collocated. Emergency management software, such as E-Team or Web EOC, allow for the instantaneous collection of incident data, resource tracking, and communications.

The decision on what layout or combination of layouts will be used in an EOC depends on a variety of factors, including available funding and floor space, expected occupancy, etc. However, there is one constant in EOC design. No matter how it is built, it will be too small. Most EOCs are not truly designed with disasters in mind. The designers usually fail to take into account the large number of personnel and outside agencies that will converge on the facility. They also fail to consider emergent functions that will require more space. For example, most designers do not plan for media operations. They may allow for a briefing area and consider seats for one or two public information officers, but they do not grasp the scope of a full joint information center that incorporates a support staff and media representatives.

The same holds true for operating requirements. The sheltering function can have representatives from a number of volunteer and public agencies and coordination of a large sheltering operation can require hundreds of support staff. EOCs simply cannot be built large enough to accommodate all the necessary functions.

This suggests two concepts. First, EOCs are coordination points where representatives agree to objectives and commit resources based on a coordinated incident action plan. The EOC is not necessarily the place where all the work is accomplished. Therefore, the EOC planners must consider where all this additional work will take place and whether these satellite facilities meet the same requirements for survivability as the EOC.

Secondly, in a large event, it may be necessary to view the EOC as the place where one initiates operations before relocating to a larger facility. New York's experience after September 11 demonstrates this concept. The initial EOC was destroyed during the World Trade Center attacks and the city had no plans for an alternate site. However, it had reserved a local pier for a large exercise and was able to occupy this site and establish a joint operations center with federal and state officials. A large event that results in a large influx of outside personnel will rapidly overwhelm the local EOC and force relocation to a larger facility. The possibility of such relocation should be part of the planning for any EOC.

The issue of relocation should also be considered when one considers alternate facilities. Most alternate sites are austere and lack the resources of the main EOC. Planning should consider whether these resources would be provided after the fact to the alternate facility or whether the alternate will provide temporary workspace while a larger facility is established.

When one considers the issue of relocating the EOC, the advantage of some type of virtual capability becomes apparent. Using Internet-based software means that relocated staff need only have a computer and Internet access to reestablish operations rather than waiting for an entire support system to be put in place. It also means that switching to mobile command posts and back to fixed sites becomes extremely easy. The point is that thinking about what constitutes an EOC needs to be expanded beyond the traditional facility to encompass all the mechanisms and facilities used to coordinate disaster activities.

One other planning parameter that is frequently overlooked for EOCs is the need to plan for sustained operations. Most EOCs are activated for only short periods of time and few are ever open long enough to establish an operational rhythm. Officials can force themselves to keep going for a day or two to meet operational needs. Disasters are different: the EOC will be open for days, weeks and even months and there is a need to plan for shift changes and long range staffing. Studies on sleep deprivation suggest that there is a marked drop in efficiency after twelve 12 hours on shift. Considering the level of stress generated in an EOC, forcing key staff to get adequate rest is essential.

COMMUNICATIONS AND INTEROPERABILITY

Crisis communications guru Art Botterell has, over the years, developed four laws of emergency management to explain what happens during emergency operations. His Second Law of Emergency Management states: "The problem is at the input." This is Botterell's way of pointing out that emergency planners often tend to view communications failures as technology problems when they are in fact caused by human factors. For example, failure to warn the public can frequently be traced back to an initial hesitation on the part of officials who seek more information before issuing a warning or who don't want to create "public panic." When challenged after the fact, it is not uncommon to blame the lack of warning on a failure of the system rather than on the hesitation.

To help explain why technology is so frequently made the scapegoat, Botterell has developed the model shown in Figure 9.13 which identifies four layers of communications:

- Organization—the structures, goals, objectives, and metrics that define the organization.
- People—human factors such as capabilities, training, attitudes, etc.
- Procedures —patterns of interaction and problem solving.
- Technology—the actual equipment and systems used to communicate.

Problems, or the perception of problems, tend to propagate downward through the four layers: a system failure is usually blamed first on technology and not on the other layers of communications. This means that the usual way of "solving" the problem is to purchase new technology. However, change tends to propagate up through the stack. Technology changes faster than the other layers, driving a demand for new procedures, skills, and behaviors to take full advantage of the new technology. The emphasis on technological rather than organizational change creates a situation where organizations neither solve the problem nor take full advantage of new technologies.

An example illustrates this point. A major city experienced a shooting incident in a high-rise building that was poorly handled. Among the issues identified in the after action report were that information

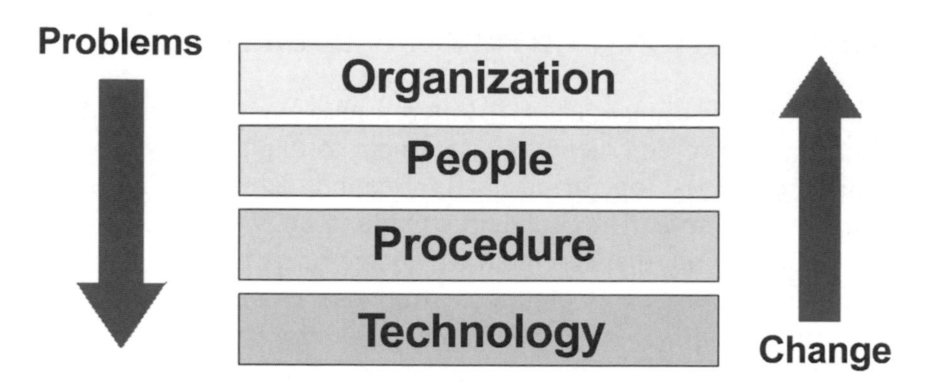

Figure 9.13 Communications model showing the relationship between technological changes and the perception of problems. Courtesy of Art Botterell.

available to police dispatchers was not made available to officers on the scene and that the existing radio system did not operate well within the high-rise building. To solve the problem, the city built a new dispatch center and purchased an advanced 800 MHz radio system. However, a closer reading of the after action report suggests that the problem was on-scene coordination (ICS was not being used as was required under city policies) and inadequate dispatcher training for critical incidents. Neither of these two issues was addressed in the jurisdiction's solution. Further, the new 800 MHz radio system was configured to resemble the old system and procedures remained unchanged, completely negating the advantages of the new system.

This example highlights the problems involved in interoperability. Since September 11, there has been a major push for making public safety radios compatible. There has been no concurrent push, however, for training responders on the deployment of radio nets, the use of liaison officers, or the development of the joint communications plans available for years under ICS. It is unrealistic to expect that all responders to an incident will operate on the same frequency, so one would expect at least some discussion of the use of radio nets such as are used in the military and the radio amateur community on a regular basis. Further, even where compatible systems exist, variations in organizational culture may preclude communications. For example, a recent controversial decision by the Secretary of Homeland Security allowed law enforcement to continue to use 10-codes, despite a push by DHS as part of NIMS to promote the use of clear text in radio communications. Since there is no standard 10-code in use in the United States and the system is only used by law enforcement, this raises a potential barrier to interoperability in disasters.

INCIDENT RADIO COMMUNICATIONS PLAN	1. Incident Name	2. Date/Time Prepared	3. Operational Period Date/Time		
4. Basic Radio Channel Utilization					
Radio Type/Cache	Channel	Function	Frequency/Tone	Assignment	Remarks
5. Prepared by (Communications Unit)					

Figure 9.14 ICS Form 205 Incident Radio Communications Plan. Although designed for radio use at the operational level, this form is a good starting point for developing a detailed tactical communications plan.

If the issues related to interoperability are not technological but cultural, then it follows that these issues could be anticipated, planned for, and resolved. When a hijacked plane crashed into the Pentagon on September 11, responding units from Arlington County and surrounding areas had more than a compatible 800 MHz radio system. They had a well-thought out and exercised mutual aid communications plan that successfully coordinated their activities. The plan included provision for integrating outside agencies into the communications system.

A major function in a crisis is the establishment of a joint communications plan that integrates existing and emergent organizations. A starting point for such a plan is completing a listing of the various systems in use, for example an ICS Form 205. While this seems simple and intuitive, it is amazing how infrequently this is done. However, knowing what is available allows for integration either through technology or by something as simple as reprogramming radios. Alternative methods of communications such as cellular and satellite phones have their place within the plan and can be provided quickly by commercial vendors. Essentially, the plan should answer the questions: who needs to talk with whom and for what purpose and how will this communication be accomplished?

INFORMATION PROCESSING

A major problem in disaster operations is information management. Day-to-day systems for managing information cannot cope with the large influx of information generated by disaster activity and the types of information normally collected may not be meet the requirement for situational awareness in a crisis. Compounding this problem is that, while many other functions in a crisis mirror day-to-day jobs, there are few employees in local government who have the job of collecting, collating, and assessing information. This makes it difficult to recruit and train personnel for the plans unit. This means that the functions of the plans unit are not adequately performed, with a concurrent impact on the flow of information within the crisis management group. An effective plans unit performs two critical functions. First, it collects and analyzes disaster information to assist in decision making and second, it facilitates development of the incident action plan.

It is important to keep in mind that information is collected for a purpose, not just because it is available. There is a tremendous amount of information available through a variety of sources and not all of it is necessary. Information is primarily collected to support decision making. This means that plans unit staff must have an idea of what is needed, how to obtain it, and what constitutes a credible source. This last point is important in determining how conflicts in information will be handled. It will also be necessary to determine what sources will be considered "official." For example, is the count of deaths based on a summary of separate incident reports or based on a daily medical examiner's report?

Information collection requires a plan; a fact that is frequently overlooked by novice plans unit members. This information collection plan should address the type of information required, potential sources for the data, the method and frequency of collection, and responsibility for gathering the information. The plans unit should have the ability to task other elements of the EOC to provide information and to deploy information collection methodologies such as aerial reconnaissance and field observers. There should be provisions for both routine collection of information and for an expedited response to a specific request.

Raw information is not useful until it has been processed. Processing means cross checking with other sources. For example, a sudden dramatic drop in shelter occupancy might mean that victims have found temporary housing or it could mean that not all shelters reported their numbers correctly. The information analyst needs to have a healthy dose of skepticism and to apply common sense. Once information has been processed, it can be disseminated through graphical displays, included in reports, or added to a database.

Facilitating the action plan, the second main function of the plans unit, involves the establishment of a planning cycle that provides for input from the various agencies involved in the plan. An incident action plan is not as difficult to assemble as one might think if one uses the ICS system. Essentially, the leadership sets objectives, each element prepares its own plan based on those objectives, conflicts are resolved at a facilitated planning meeting, and the plans unit assembles the final document. The secret behind the process is the facilitated planning meeting, which focuses on conflicts and problems rather than an information exchange.

By this point, this process should sound familiar. The incident action plan is based on management by objectives and uses the same process as the strategic plan and work plans discussed in previous chapters. The management team develops the performance objectives for the next operational period. An action agent is identified, usually through a cooperative process at the planning meeting. The objectives may also be broken down into smaller tasks based on the contributions of other agencies. For example, if the performance objective is to deliver X gallons of water within the next 24-hour period to a specified location, this task may be given to the water department. However, the water department may also break it into smaller tasks that involve purchasing bottled water, arranging for the loan of water trucks from the National Guard, and establishing a water distribution point using a federal water purification unit.

MUTUAL AID

By definition, a disaster overwhelms local resources. The generally accepted strategy for overcoming this shortage of resources is to seek assistance from other jurisdictions in the form of mutual aid. Mutual aid is a pre-arranged agreement, which may or may not have a financial component, to provide essential resources when local resources are inadequate to meet the needs of a disaster. While the concept is simple, there is significant disparity in how the concept is applied across the country.

At one end of the scale are states like California and Florida that have extensive and robust mutual aid systems. California has a master mutual aid agreement signed by all counties that provides for the cost-free provision of needed resources in an emergency and establishes safeguards to prevent a jurisdiction from being stripped of all its resources. California has separate mutual aid agreements for disciplines such as fire, law enforcement, and emergency managers that provide for financial reimbursement. Florida uses as system in which a jurisdiction queries a statewide system for resources and receives a list of available resources and cost estimates. The jurisdiction then selects the most cost-effective resource.

At the other end of the spectrum are states that have little or no real mutual aid. In many cases, what little mutual aid is available is limited to fire services. Some states do not recognize the authority of peace officers outside their home county and the sheriff of the requesting county must swear in mutual aid personnel before they can be deployed. The problem is particularly acute in states that have home rule, with the autonomy of multiple small jurisdictions making it difficult to implement a statewide system.

Even where mutual aid systems exist, not all the details of reimbursement of costs, liability, and workman's compensation have been worked out. For example, there is no national credentialing system for medical resources provided under the Emergency Management Assistance Compact (EMAC), the system that provides state-to-state mutual aid. Reimbursement for resources deployed under EMAC has not yet been addressed in federal regulations. Consequently, one of the considerations in requesting and deploying mutual aid is the existence of agreements that address such issues. These are not they types of issues that can be quickly resolved during a crisis.

A further consideration for mutual aid resources is the need to support them. Jurisdictions with limited experience in deploying mutual aid resources sometimes fail to plan for the type of sustained support needed in a major crisis. Mutual aid resources provided for routine emergencies are generally used for short periods and are dispatched enroute. This means they are dispatched directly to the scene and are essentially self-sustained. In a major disaster, mutual aid resources are deployed for longer periods of time and must be supported. Initial deployment may be to a staging area where the resource is prepared for assignment to an incident (e.g. briefed, refueled, rested). For such long deployments, mutual aid resources will need to be housed and fed during rest periods, may need maintenance and fuel, or may need medical support. For this reason, while operational elements may request and deploy mutual aid resources, there is a tactical component to mutual aid and such deployments must be supported through the EOC.

RESOURCE MANAGEMENT AND LOGISTICS

Disasters are ultimately about resources. This is not to diminish the importance of operational response. Rather it is an acknowledgement that jurisdictions are staffed to provide services at a level that meets daily needs. A disaster, by definition, overwhelms local resources, including those in the public safety arena. The success or failure of operational response will depend largely on the ability of the jurisdiction to obtain and deploy outside resources. Yet resource management is a major point of failure in many crises.

Why is this so? Aside from seeming intuitive, every major reference in emergency management—SLG 101, NFPA 1600, the EMAP Standard, NIMS, the NRP—highlights the importance of resource management. Yet the NEMB-CAP study found only a 17 percent conformance with the EMAP Standard for resource management and 29 percent conformance for the logistics standard. Further, jurisdictions are frequently criticized during crisis for failing to obtain or distribute essential resources such as food and water, as witnessed at the Superdome in New Orleans during Hurricane Katrina. Clearly something is not working.

A very big part of the problem is that, for most jurisdictions, there is confusion over the definition of resource management and an unrealistic expectation that normal procurement procedures will be sufficient to manage a crisis. In essence, the jurisdiction confuses procurement with logistics.

Resource management in the context of NFPA 1600 is broad in scope. It refers to any resource that is needed to support overall emergency management program goals and objectives. The intent of the standard is to establish a baseline, determine shortfalls, and develop plans for meeting those shortfalls such as through procurement or developing mutual aid agreements. In this context, a resource may be tangible, such as a piece of equipment or a specialized team, or it may be something intangible such as training or funding.

This broad definition in NFPA 1600 is somewhat confused by SLG 101's use of the term "resource management" to refer to its logistics annex. Since the NRP uses the term "resource support," one hopes that anticipated revision of SLG 101 will bring its terminology more in line with current practice. For the purpose of this discussion, this book uses resource management in its larger context as a strategic concept and not as a specific emergency management function.

Where resource management is a broad planning concept that addresses multiple types of resources, logistics is a very specific function. It is the process by which a jurisdiction obtains and distributes needed resources. This involves a complex network that allows for a requestor to order a resource, track its status until received, or know that it is unavailable. This implies a number of functions:

- Accepting the request
- Validating the request
- Determining how the order will be filled
- Placing the order
- Tracking the order
- Warehousing and transshipping
- Receiving the order
- Verifying delivery
- Paying for the order
- Tracking costs
- Recovering durable property.

Disaster logistics may require the establishment of new facilities such as warehouses and staging areas. It may require developing distribution points to provide goods directly to the user. For example, the State of Florida has an extensive distribution plan that moves goods through a series of warehouses, marshalling areas, and staging areas to points of distribution (PODs) established by the local jurisdiction. The plan specifies responsibilities for each level of government and describes the local requirement for developing POD sites. Logisticians may also find themselves dealing with requests that are not normally the task of local government, such as obtaining lodging for incoming mutual aid personnel. The FEMA team during the Northridge earthquake had to find a circus tent within 24 hours to fulfill a commitment to local merchants.

Procurement, the task that most local governments do well, is the ordering of goods or services through a contracting process. This is a very small component of an overall logistics function. Unfortunately, the concern over protecting public funds from waste, fraud, and abuse makes most of these systems too cumbersome for use in a crisis. For this reason, most emergency authorities allow for waiving procurement requirements. However, the tracking systems that are part of this process are vital in recovering funds under insurance or federal recovery programs, a fact that is frequently overlooked. Just because a jurisdiction faces a crisis does not mean that accountability can be disregarded.

Clearly, attempting to develop an ad hoc logistics system at the time of crisis is not a good thing, but it has been done. On September 11, a Long Island police officer reported to New York under mutual aid and was given the assignment of creating a logistics system from the ground up. The process was painful, to say the least, and it took several weeks to reach a reasonable degree of efficiency. If one elevates this to the level of catastrophe, it becomes obvious that attempting to create a logistics system without any pre-planning is a recipe for failure.

As part of this pre-planning, it is important to involve existing procurement and financial staff. By tracing the logistics process from initial request to final accountability, it is possible to identify barriers, such as needed changes in procedures, and required authorities. The establishment of a base camp for firefighters provided under mutual aid was delayed for several weeks during Hurricane Katrina because no one had the authority to commit the funds at the required level, even though federal reimbursement was guaranteed. Again, a logistics program is not just about delivering goods and services; it includes the financial accountability that is expected of local government.

One potential problem in logistic planning is the NRP operational structure being adopted by most states. The structure calls for a resource support function (ESF 7) but does not say where that function resides. The organization chart reflects a resource management branch as part of the logistics section. However, the logistics management support annex assigns logistics management to ESF 5 Emergency Management and describes a process where state requests are initially sent to the operations section and passed through ESF 5 to logistics if the resource is not available in a staging area. ESF 7 is not mentioned in the process. The ESF 5 annex states that ESF 5 provides staff to the logistics section leader for a list of tasks that encompasses all logistics functions and requires the logistic section chief to work closely with ESF 7 and implement the logistics management support annex.

The confusion begins to make some sense when one considers what is actually going on. A request from the state is sent to the operations section where it may be filled from resources in a staging area under the control of the operations section chief. These represent the most immediately available resources. If the resources are not staged, the request passes through ESF 5 for some arcane federal purpose to the logistics section where it may be filled from resources located at a logistics base under the control of the logistics section chief. If the resource is not available, the logistics section chief can mission-assign a federal agency to provide it, pass the request to the national level, or purchase the resource, with ESF 7 serving as the procurement unit.

The point of this convoluted treatment is that blindly adopting the NRP operating structure is not necessarily in the best interests of a local jurisdiction. The process described in the NRP may work for the federal government but it is most likely too cumbersome for a local jurisdiction. Establishing a strong logistics unit along the standard ICS lines may considerably streamline this process for local government and there may be no need for ESF 7 (or for that matter, ESF 5) in local plans. As was discussed in Chapter 8, plans must realistically address local needs and not simply conform to a format. The NRP operating structure is not mandatory for any level of government.

The discussion on the NRP also calls into question the need for an operations section at higher levels of response. Since tactical-level structures such as MAC groups deal primarily with resource requests, the structure described in Figure 9.8 might actually reduce some of the confusion between what is handled in operations and what is handled in logistics.

THE JOINT INFORMATION CENTER

Crisis communications consultant Art Botterell has formulated a Fourth Law of Emergency Management that states: "Expectation is reality." This reflects the fact that an operation is not necessarily judged by what it does but rather by what it appears to have done. In every crisis there is a defining moment. If that moment is handled well, the response to the crisis will be judged as successful. If handled poorly, there is no retrieving the damage to the community's reputation.

Consider the situation at the Superdome in New Orleans during Hurricane Katrina. Relief reached the victims in just over three days, consistent with the standard emergency management guidance that all citizens be prepared to be self-sufficient for three days in a disaster. There is also evidence to suggest that food and water was available at the Superdome and media reports of violent crime and death have been found to be exaggerated or inaccurate. Yet the single most memorable image of the disaster is the line of victims chanting for help before television cameras, and the failure of the federal government to provide immediate and overwhelming assistance to the site was a public relations disaster from which FEMA never recovered.

The surprising thing about the Hurricane Katrina operation was how badly FEMA handled the media, as FEMA was the originator of the highly effective Joint Information Center concept. The JIC is a mechanism to maximize the use of public information resources, identify a single point of release for information, and provide the media with a source of reliable data and fact checking.

An important consideration in selecting a lead public information officer is that there is a considerable difference between the day-to-day duties of a public affairs officer and the role of a crisis communicator. With media networks running continuous coverage of high-visibility disasters, information demands in a crisis are much greater than normal and there is no longer any concept of deadlines. News is immediate and requires constant refreshing. This means that there must be a close link between the JIC and the plans unit to ensure that information is timely and accurate and a streamlined process for deciding what information can be released and what must be withheld.

The core of the JIC consists of the PIOs from all the agencies involved in the response operation. These PIOs meet to exchange information, resolve conflicts, and develop a "message." The "message" represents the official view of the JIC in terms of statistics and instructions to the public. This "message" can be released through all the normal media outlets, is disseminated to field PIOs, and is used as a script by staff manning a media phone bank. Calls that cannot be answered from the script are transferred to the appropriate agency PIO.

A fully functional JIC will normally have workspace for media representatives and a permanent briefing room with an appropriate backdrop, such as visual displays or the phone bank staff. In essence, this is a set and care must be taken to make sure staff realize that anything said or done on the set is accessible to the media. A full JIC will normally include the capacity to monitor broadcasts to ensure that the media is putting out accurate information or to identify potential problem stories. It will also include the capability for arranging for press credentials, interviews, background footage, or any other media needs.

The JIC represents a significant expenditure of space and resources. However, it is the single best way to deal with the large influx of media that will descend on a disaster scene. The JIC cannot control the news, but it can ensure access to accurate information. It maximizes the use of the PIOs, ensures that consistent information is being put out at all levels, and provides the press with access to decision makers. More importantly, it depicts the disaster management organizations as organized and proactive.

CONCLUSION

Tactical response—the coordination of the resources involved in the disaster—provides the support needed to sustain the operational response and to transition to long-term operations. Successful response requires an understanding of the dynamics of disaster operations and the realities of community response, aggressiveness in providing victim services, and creativity and flexibility in problem solving. Pre-planning is essential, but ultimately it is the ability to think beyond the limits of the pre-planning and to adapt to a changing environment that will determine the success of the disaster response.

Chapter 10

MANAGING CRISIS

Expectations are often premeditated resentments.
—Jonathan Bernstein
Crisis Management Consultant

Emergency managers tend to focus their efforts on tactical planning. The early chapters of this book explored some of the reasons for this, primarily the origin of emergency management in the civil defense program. It is an area where emergency managers feel comfortable. However, successful disaster management requires a response on all three levels: operational, tactical, and strategic. It is a sad fact that the perceived success or failure of a response is not always determined by what is actually done, but by how well public expectations are met.

When one compares immediate response to short and long term recovery, it is obvious that response is only an initial stage of relatively short duration. Because it is an important stage, most emergency preparedness activities are focused on building response capacity. Nevertheless, recovery and reconstruction efforts ultimately determine the continued viability of a community and the length of time needed to return to some acceptable state of normalcy.

It is in these two areas of public expectation and perception and recovery and reconstruction that strategic response becomes important. Unfortunately, most emergency plans do not address strategic response beyond the formation of a "policy group" or consideration of federal recovery programs. There is a failure to recognize the need for organizing for community recovery in the same way that response is organized. Preparedness efforts must include the establishment of plans and mechanisms for strategic response. Unlike tactical response, there are few emergency management sources for preparing for strategic response. However, there is a considerable body of literature on crisis management that has relevance to strategic disaster response.

LEADERSHIP AND DECISION-MAKING

Social science has demonstrated that community leadership is generally not fully engaged in emergency planning. This means that, while senior officials may be supportive of the emergency management program, they are unlikely to commit the time necessary to provide direct leadership to the program or to participate in exercises or training. Yet these same officials are expected to provide leadership in a crisis. Unfortunately, most of this leadership translates into trying to direct response from the EOC as opposed to focusing on strategic issues.

Figure 10.1 summarizes the four Laws of Emergency Management formulated by crisis communications consultant Art Botterell. While other chapters have made mention of these laws, it is in the role of senior officials that all of the four factors come into play. Key government officials are usually untrained in their roles. This creates misperceptions based on disaster mythology and creates stress, leading in many cases to hesitation in decision making.

In *Crisis Management: Planning for the Inevitable*, Steven Fink discusses the impact of stress on decision making and points out that stress is both a negative and a positive. A certain amount of stress engages an individual. This, after all, is the biological origin of stress —to prepare an individual to deal with crisis either through aggression or by flight. Too much stress, however, reduces an individual's ability to think clearly and proves the truth of Botterell's first law: stress makes you stupid.

Botterell's Laws of Emergency Management

1. Stress makes you stupid
2. The problem is at the input
3. No matter who you train, someone else will show up
4. Expectation is reality

Figure 10.1 Botterell's Laws of Emergency Management. Courtesy of Art Botterell.

Fink confirms this by identifying four potential decision-making mechanisms, only one of which is viable in a crisis:

1. Unconflicted Inertia or Unconflicted Adherence—this is the tendency of a decision-maker to cope with crisis by ignoring relevant information and continuing to perform "business as usual." This can be reflected by a rigid adherence to a plan or a set course of action with no attempt at innovation or creative problem solving.
2. Unconflicted Change—indecisive leaders have a tendency to follow the last advice given to them. This creates the impression of vacillation and raises questions about the official's leadership. This mechanism can be seen regularly as politicians change unpopular positions on the basis of media attention or public outrage rather than facts.
3. Hypervigilance—this is the most maladaptive of the decision-making mechanisms and, unfortunately, is one of the most common in disasters. Stress levels exceed the point where they are useful and, in a near panic, officials attempt to resolve the crisis

in any way possible. This is the manic behavior seen in EOCs when officials who are not familiar with emergency plans and who have never attended an exercise seek to influence tactical decisions without regard to expert advice.

4. Vigilance—unlike the previous three mechanisms, vigilance is adaptive behavior that encourages the collection of relevant facts, an objective assessment of those facts, and the articulation of a well-reasoned decision.

In considering these decision-making mechanisms, one is appalled by the implication that there is only a 20 percent chance that a senior official will get it right. This emphasizes the need to prepare senior officials for decision-making that goes beyond the usual tabletops related to response. Deciding on curfews and reentry of evacuated areas are well within the capacity of the tactical planners at the EOC. Instead, one must consider how to prepare managers to cope with stress and to focus on long-term issues. Based on this understanding, it is possible to derive three corollaries to Botterell's Laws of Emergency Management that can assist officials during crisis:

1. Don't just do something; stand there! Most senior officials do not understand the pacing of tactical-level functions. There is a tendency to "do something" and a failure to realize that response is already taking place on the operational level. The first step in any crisis is to pause and ask, "What do we know?" Doing "something" immediately is not managing the crisis. Craig Fugate, director of the Florida Division of Emergency Management suggests, "Take a deep breath and then your pulse. As long as you have both, relax." Relaxation techniques such as these can help reduce stress to manageable levels.
2. Don't try to fix the world; fix the problem. General Dwight Eisenhower once said, "The older I get, the more wisdom I find in the ancient rule of taking first things first, a process which often reduces the most complex human problem to a manageable proportion." Disasters are, by definition, overwhelming. It is not possible to hit them head on and survive. Instead, the manager's job is to analyze need and to prioritize response based on existing

resources. Not everything needs to be fixed immediately. This concept lies at the heart of continuity planning and the development of recovery time objectives. It works for emergency operations as well. Food is not as important as water. Disposal of the dead is not as important as emergency medical services. Firefighting trumps debris clearance. This corollary speaks to both the reduction of stress and the acquisition of relevant data through limiting problems to manageable levels.

3. Truth, no matter how harsh, is a planning parameter. Accurate information is the basis of good decisions. Senior officials can unwittingly create environments where subordinates become unwilling to reveal shortcomings or perceived failings. This is particularly true when the official has unrealistic expectations for the pace of operations. Yet the nature of disaster operations is that problems can be solved a variety of unconventional ways. First, though, one must acknowledge that there is a problem. Another way of considering this is to "fix the problem; not the blame," a concept alien to many U.S. politicians. Having relevant data and, more importantly, confidence in the data can further reduce the stress associated with decision making.

Dr. E.L. Quarantelli identifies a number of common problems with decision making at the EOC. The first problem is the tendency of senior leaders to want to remain in the EOC for extended periods of time. Part of this is, of course, the unwillingness to delegate authority for a career-breaking crisis to an ad hoc organization. This uncertainty is the direct result of not having participated in exercises or smaller emergencies and consequently having little trust in the system. However, as was noted in the previous chapter, disaster operations require sustained operations. The longer an official goes without adequate rest, the closer he or she is to collapse. When this occurs, particularly if the executive has reserved all decision-making authorities to himself or herself, the EOC may find itself without the authority necessary to continue vital operations such as purchasing relief supplies or approving mutual aid requests.

One of the characteristics of a disaster is the creation of new tasks that are outside the normal day-to-day responsibilities of the jurisdictional organization. Major policy issues may arise out of the need to assign responsibility for these tasks. Existing organizations may be reluctant to take on new tasks. There may be insufficient legal authority to accomplish the new tasks. There may be conflicts over jurisdictional issues. These are strategic issues that require the in depth understanding of community politics and group interaction that are the hallmarks of a good politician. They may also require the imposition of political will, emergency legislation, or a policy decision to ensure the integration of organizations and tasks.

Disasters also result in the emergence of new or substantially changed organizations. Figure 10.2 is a model of group adaptation in disasters based on the work of Dr. Russell Dynes and others. The model considers changes to the structure of the organization and to the type of tasks performed. Type I organizations remain unchanged during disasters, keeping the same structure and performing the same tasks. Type II organizations increase in size and may change in structure during disaster but perform the same tasks normally performed pre-disaster. Type III organizations keep the same organizational structure but take on additional tasks during emergencies. Type IV organizations are newly formed groups that were not part of the pre-disaster community.

Type IV organizations carry the most potential for creating problems if not appropriately integrated into the disaster management organization. They may be community-based or external to the community and may be viewed with suspicion by established organizations. This is particularly true if the emergent organization is perceived as encroaching on the responsibilities of the established organizations or is viewed as a competitor for resources or funding. This situation is not uncommon among volunteer organizations. However, not all emergent groups represent the community and providing official recognition to such groups may be inappropriate. Consequently, the decision to integrate newly emerged groups into the disaster management organization may need to be made at the strategic policy level.

No area is more fraught with controversy than issues arising over jurisdiction. Disasters have a way of straddling both geographic and jurisdictional boundaries. During crisis, these overlapping responsibilities and authorities represent a significant potential for conflict.

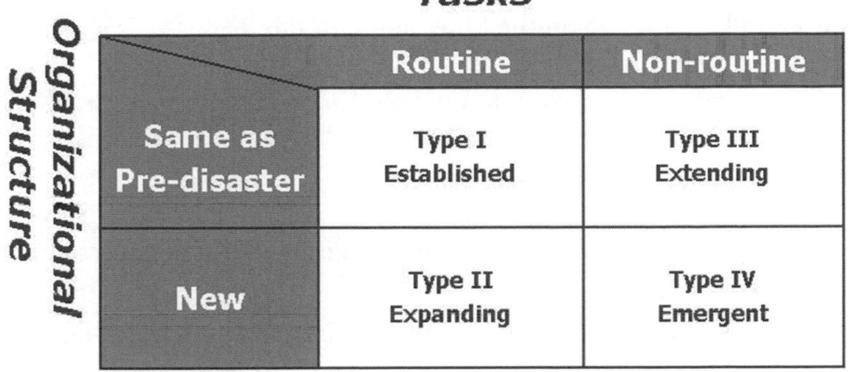

Figure 10.2 Disaster Research Center Model—Typology of Organizational Adaptation in Crisis. From *Facing the Unexpected: Disaster Preparedness and Response in the United States.*

However, there are mechanisms to help resolve these issues, such as the use of unified command, and senior officials should bc quick to identify and diffuse such conflicts. Many of these issues have a political dimension and are related to public policy and so must be decided by the political leadership of the jurisdiction rather than by a response organization like a MAC Group.

As one considers the decision-making mechanisms needed in a disaster, it becomes apparent that the skill sets needed in disasters are different than that used in day-to-day activities. Executives are used to making decisions based on the known variables encountered on a daily basis. A disaster, however, is characterized by uncertainty. Decision making based on trial and error becomes more the norm than decision making based on a systemic approach (i.e. one that replicates a method known to lead to a solution). This places a premium on collaboration and consultation rather than direction and control. In other words, decisions are made by drawing on the expertise of many as opposed to being made on the knowledge of the leader.

This changes the nature of the organization needed in disaster response from the mechanistic one used in day-to-day operations to one that researchers Burns and Stalker term "organic." Organic organizations are characterized by:

- Organizational structures that provide for task force approaches to the division of labor.
- Task assignments that are not rigidly defined and adjustable based on the changing situation.
- A communications structure that emphasizes the flow of information.
- Emphasis on coordination rather than on command and control.
- Emphasis on self-control and discretion rather than on system control.
- Limited preoccupation with adhering to the chain of command.

This concept of an organic organization speaks to the strengths of the multi-agency coordination system discussed in Chapter 9. It reinforces the idea that the EOC operating structure should be based on MACS and not the field ICS structure and that the appropriate role for the EOC is one of coordination rather than command and control.

If one accepts the idea that the EOC is a point of coordination and not of command, then it follows that the senior leadership of the community does not have to be present in the EOC at all times to direct operations. During the initial phase of a crisis, it is normal and appropriate for senior managers and department heads to meet at the EOC, receive a briefing on the crisis, and set policy. However, as the disaster progresses, the appropriate role for these key leaders is directing and controlling the field operations of their department, not overseeing activities in the EOC. Ideally, senior leaders will meet once or twice a day to approve incident action plans or to resolve policy issues and spend the rest of their time doing their jobs.

This is not to suggest that senior officials and department heads have no role in disaster response. On the contrary, their roles are vital. However, the traditional model of response that absents a department head from their department at the most critical moment of their career is counter-intuitive. This is precisely the time that departments need decisive leadership and the presence of the department head is critical to the success of operational response. Further, once one delegates tactical decisions to the disaster management team or MAC group, the level of decision making required of senior leaders drops dramatically. As long as the disaster management team has the authority needed to obtain and allocate resources (e.g. fiscal authority, authority to request mutual aid, etc.), there should be few tactical decisions requiring the attention of senior department heads, freeing them to concentrate on directing their departments' operational response.

Allowing oneself to be distracted by tactical or operational issues is common, even among professionals. FEMA deployed its National Emergency Response Team to provide relief following a hurricane in the Caribbean. The initial team elements landed on an unaffected island in another state and the federal coordinating officer and key members of his staff went to the main affected island to coordinate with the local government. While there, the FCO became involved in assisting with the evacuation of tourists from the island, and it was two days before he was able to return to the staging area and direct deployment of his team, which, at this point, had grown to several hundred workers. Because he had been focused on operational rather than strategic issues, the deployment was sporadic and no work had been done to determine operating facilities, transportation, or lodging for the team. Because of the press of time, strategic options, such as splitting the team into a rear base and a forward team to reduce the competition for housing on the affected island, were not even considered.

CRISIS MANAGEMENT

So, if senior executives are not managing response, just what do they do? The answer is that they manage the crisis, not the disaster. This is more than semantic double talk. The disaster is immediate: people must be rescued, shelters must be established, and critical infrastructure must be restored. However, these are tactical and operational issues that will be sorted out one way or another. Eventually, enough resources will be mustered and deployed to solve these problems. The real crisis is the long-term impact of the disaster on the community. For example, following the earthquake in Kobe, Japan, several large companies took the opportunity to relocate to Malaysia where labor costs were considerably cheaper. Following Hurricanes Katrina and Rita, the population in New Orleans has dropped from 484,674 to fewer than 150,000. Obviously, the relocation of major employers and a reduction in population have a drastic impact on the tax base and the community's ability to fund recovery projects.

As counter-intuitive as it may seem, the principal focus of the senior officials of the jurisdiction is not the immediate response problems but the identification of long term impacts and the establishment of a strategy to deal with those impacts. It is on public policy and political issues. It is on community resilience. So, the first step in managing crisis is to properly identify the actual crisis. The crisis is not always obvious and immediate needs are sometimes only the indicators that point to the actual crisis. The senior official needs to ask, "What's really going on here and what does it mean in the long-term?" In many cases, the real crisis of a disaster is not in meeting the immediate needs of the victims, but in preserving the economic viability of the community and funding long-range reconstruction. Just as in hazard analysis, one must consider impacts rather than the agent that caused them.

Complicating this issue is the fact that it is not always apparent that a disaster is occurring. Social science suggests that people tend to "normalize" emergencies. That is, they tend to view them within the context of day-to-day activities rather than recognizing that something unusual is occurring. This is part of the reason that public warning notices are not always issued in a timely fashion or why EOCs are not fully staffed in the face of potential crisis. Sometimes the difference between emergencies, disasters, and catastrophes is slim: the breaching of the New Orleans levees transformed the nature of Hurricane Katrina; conflagration following the 1906 earthquake destroyed most of San Francisco. Recognizing that a crisis is occurring and taking steps to manage it are easy in concept, but difficult in execution.

Focusing on the strategic aspects of the disaster rather than the tactical and operational ones allows isolation of the crisis, another concept in crisis management. Isolating the crisis means focusing directly on the crisis and not being distracted by other issues. It is not possible to run the operational response, manage day-to-day functions, and deal with strategic issues all at once. This means that senior officials need to consider how much they will be involved in each and to whom they will delegate other responsibilities. This can, of course, change over time. Art Agnos, the mayor of San Francisco during the Loma Prieta earthquake, was not particularly popular with the voters, but he enjoyed a surge of popularity when he devoted himself full time for several weeks to dealing with policy issues resulting from the earthquake. Similarly after September 11, Mayor Rudi Guiliani of New York devoted himself full time to recovery issues and was widely praised for his very visible leadership during the crisis.

There is a further reason for maintaining a strategic focus. Despite a perceived history of "rugged individualism" and independence, Americans tend to play by the rules. Far from being a nation of risk takers, Americans do not generally reward initiative in public service. David Brown of the Washington Post suggests that this is the result of a litigious society that raises the specter of legal liability whenever risk is considered. A second factor is a fear of criticism and public humiliation from reporters always willing to find fault with government.

However, social science research suggests that effective disaster response depends on innovation and creative ideas. Tactical responders must be willing to adapt plans to changing situations and to use resources in unexpected ways. For example, during Hurricane Iniki, urban search and rescue teams were used to evaluate the fire load created by vegetation killed by the storm and to clear sites particularly prone to fire. Since there was no requirement for their primary mission, the teams used their skills in a way that was not pre-planned to address an immediate threat. The inherent conflict between a tendency to follow the rules and the need for innovation could potentially be resolved though the vision, influence, and political will of senior government officials.

Identifying and isolating the crisis are two important first steps toward managing the crisis. They remove distractions and allow a focus on the real issues. However, at some point, senior officials have to take decisive action to manage the crisis. There is a very real danger that issues will be over-studied and no action taken, what crisis management consultant Steven Fink calls "analysis paralysis." Like tactical and operational response, strategic response requires aggressiveness and a certain amount of risk taking. As General Patton was fond of saying, "A good plan executed today is better than a perfect plan executed at some indefinite point in the future."

Planning for Crisis Management

Core membership in a crisis action team should be pre-planned. At a minimum, the team should consist of the chief elected or administrative official, legal counsel, lead PIO, chief financial officer, and key department and agency heads. This core membership can be augmented at the time of crisis as needed. It is important to remember that, while many in this core group may have operational authority, the focus of the crisis management team is on long-term issues of public policy, not the ongoing response operation. This does not preclude the team from providing appropriate oversight and quality control of the response (indeed, they would be negligent if they did not), but the resolution of any issues should be delegated to the appropriate tactical or operational organizations. If the crisis management team bypasses these organizations and begins to issue instructions directly to operational units, then these tactical and operational organizations become marginalized and coordination breaks down.

It should be obvious by now that trying to develop strategic response on the fly is difficult at best. Ideally, many of the issues and potential solutions or options could be identified during the development of recovery and mitigation strategies and incorporated into long-range plans. Issues of the types discussed above lend themselves very well to tabletop exercises. In fact, this is the preferred mechanism for training crisis management teams. As has been noted, the pacing of response varies from level to level. Operational response is characterized by the need for immediate and urgent action. Tactical response requires a step back from the operational and a focus on supporting requirements and resource allocation. The pacing for strategic response is even slower. Consequently, tying the crisis management team or policy group into a functional or full-scale exercise is not always the best way to train crisis managers. There is just too much temptation to get involved in the tactical response, as true policy issues are not necessarily immediate issues and may be slow to emerge in the initial stages of the crisis.

One advantage to recognizing the three levels of response is that the model plays to the strengths of those that implement them. Front line departments are used to dealing with the immediacy of emergencies. They know the resources available, the qualifications of their team, and the types of issues that will arise. Personnel at the EOC require training to overcome an operations-oriented mindset but generally can carry over knowledge from their operational experience that enhances their value as coordinators. Likewise, strategic response plays to the strengths of local officials. Elected officials are keenly aware of the need to build public consensus and of the types of issues that resonate with the public. They understand how to provide policy direction and how to craft legislation. They understand how to deal with other levels of government and high-ranking officials.

Given these strengths, it is amazing how many officials feel the need to become involved directly in tactical response, an area for which few are really suited. By doing so, they fail to isolate the crisis and set themselves up for failure. The role of senior officials is to identify long-range issues, set goals, craft strategies to achieve those goals, and establish the policies needed to implement the strategies. Training for senior officials should emphasize this role and consider tactical response only in the context of oversight and quality control.

Hurricane Katrina provides a powerful case study of how failure to identify and isolate crisis can have tragic results. One of the points of initial confusion was the new Homeland Security designation of "Incident of National Significance." Although President Bush had declared a state of disaster under the Stafford Act and FEMA had deployed the National Emergency Response Teams prior to landfall, Secretary Chertoff has been blamed for not declaring an "Incident of National Significance," which would have invoked certain authorities under the National Response Plan. Prior to the formation of the Department of Homeland Security, however, a declaration of disaster under the Stafford Act provided all the necessary authority needed for the federal coordinating officer to manage the activities of the federal agencies responding under the Federal Response Plan.

Further complicating the situation was the decision to appoint a principal federal official as called for in the National Response Plan and to appoint Michael Brown to the position, despite his never having been trained as a PFO. The PFO's role is ambiguous under the National Response Plan, with no real authority or staff support. In essence, the National Response Plan added a layer of bureaucracy and reduced the ability of the federal coordinating officers to deal with tactical response.

Brown immediately involved himself in tactical rather than strategic issues and became the focal point for all that was going wrong with the federal response. (There is a suggestion that the FCO may also have become too involved in tactical response. Emails suggest that he was directly coordinating the flow of relief supplies from other states rather than delegating this to his operations section.) This severely reduced his ability to effectively manage the overall crisis by drawing him almost exclusively into the tactical issues surrounding the response in New Orleans.

If one considers this situation from a crisis management perspective, one must ask, "What is the real crisis here?" The crisis was not the flooding of New Orleans or the stranding of disaster victims at the Superdome. The real crisis was the lack of coordination in the federal response. The situation in New Orleans was merely a symptom of the crisis, and focusing on the single symptom ignored the situation in the other states affected by Katrina. Many communities were in even worse condition than New Orleans.

Assuming that, as PFO, he had authority to deal with the coordination issue, Brown could have considered setting specific priorities, such as dealing immediately with the Superdome issue, and delegated their implementation to the FCO and his Emergency Response Team, who were trained and equipped to deal with tactical issues. His concern should have been to see that the FCO had the needed resources, redirecting them from other operations if necessary, and then to focus on improving overall federal response. He could not do this and directly manage the situation in New Orleans. In crisis management terms, Brown failed to isolate the crisis.

It is, of course, easy to second guess decisions made during a crisis. However, the point is that, allowing oneself to be drawn into tactical issues, no matter how pressing, sacrifices one's ability to react strategically. This is not to say that the tactical issues are not important—an overwhelming response to the Superdome crisis might have changed the public impression of the federal response in Katrina. However, this type of response should have been within the capabilities of the tactical staff at the joint field office and should not have required Brown's personal intervention. If it did, fixing the problem at the JFO would have been the appropriate focus of his efforts, not the delivery of relief services to New Orleans.

There were other problem areas that could have been addressed by focusing on strategic issues. FEMA has come under considerable criticism for its use of no-bid contracts to support reconstruction and immediate recovery. The policy decision as to how contractor support would be used could have been anticipated in the early days of the catastrophe and resolved before the first contracts were let. Similarly, the decision to scatter Katrina refugees across the country led to concerns over how long the federal government would fund living expenses. While expedient at the time, it will have a long-term effect on New Orleans if a large percentage of the evacuees choose to make new lives for themselves and do not return. Again, these are strategic issues that could have been anticipated and addressed before they reached the level of a crisis.

Effective crisis management can be enhanced by organizational change as well, something that could potentially be encouraged by emergency managers with direct access to key officials. Weicke and Sutcliffe identified five common characteristics of "high reliability organizations," such as aircraft carriers and nuclear power plants, where failure carries extreme consequences. These organizations all shared the following characteristics:

1. Preoccupation with failure—identification of problems or mistakes was rewarded rather than punished. This sensitized the employees to the slightest errors that could have potential long-range consequences. If one looks at this in the emergency management context, reacting to potential crisis by activating the EOC or deploying resources would be viewed as proactive rather than over-reactive. However, if no crisis occurs, emergency managers are frequently accused of overreacting and wasting time and resources.

2. Reluctance to simplify—a major way of simplifying things is to make assumptions. However, assumptions can alter expectations and distort what is actually happening. The lesson for emergency managers is to base programs on hard information such as social science or hazard studies rather than on general assumptions and conventional wisdom. This helps to avoid basing plans on disaster myths and helps to encourage the development of realistic and adaptable options for response.
3. Sensitivity to operations—a phrase that is surfacing with regularity in the finger pointing over Hurricane Katrina is "situational awareness." This is the ability to see the big picture, to understand the dynamics of the crisis as a whole, and to see where individual activities fit in. This is precisely what should be taking place at the tactical and strategic levels of response. Rather than focusing on the operational, the MAC Group or emergency management team at the EOC tries to anticipate problems and support requirements; the crisis management team is doing the same thing for long-range issues.
4. Commitment to resilience—resilience, in this case, is to understand what has occurred and to react to it before it becomes worse. In other words, fix the problem, not the blame. Intelligent reaction, improvisation, and creative problem solving are the characteristics needed. One must be able to react quickly to a crisis, first by recognizing that one has occurred and then by implementing response options appropriate to the crisis.
5. Deference to expertise—in hierarchical organizations, deference is always accorded to those higher up in the structure. However, these are not always the individuals who have the knowledge or expertise to manage crisis. High reliability organizations have recognized this and allow decisions to be made by those with the appropriate knowledge and expertise. The emergency management equivalent of this is the use of MACS. This allows input from experts of all types in a collaborative problem solving process. This concept also comes into play when members of the crisis management team allow the tactical team to function without interference and the tactical team lets the field operational commanders handle their assignments without intervention.

Re-orienting emergency management programs to take into account these success factors could go a long way to improving a jurisdiction's capacity to respond. This is particularly true in terms of strategic response. Building a high reliability organization increases confidence levels during crisis, thereby reducing the stress on senior officials. Further, it provides for the resolution of many problems before they reach crisis proportions.

CRISIS COMMUNICATIONS

Chapter 9 briefly discussed the role of the Joint Information Center (JIC), a mechanism designed to centralize media operations and ensure a unity in the jurisdiction's message. In the context of tactical operations, the bulk of this message relates to information needed by the public to deal with the crisis directly. Examples of this type of information are hazard warnings, shelter locations, evacuation routes, etc. In the strategic context, however, crisis communications is more about the message and symbolism than it is about hard information.

To a certain extent, crisis communication is based on public expectation and the perception of how well these expectations are being met by local government. Botterell's fourth law is quite correct—public expectation does become reality, even if only in a symbolic way. FEMA became a response agency not because of any strategic plan on the part of its leaders, but because the public expected it after Hurricanes Hugo and Andrew. For this reason, public officials have been setting themselves up for failure for years by raising expectations among citizens that government could respond quickly and effectively to overwhelming crisis even though it lacks the resources to respond to day-to-day emergencies. If public health systems are overwhelmed by the annual flu season, can one expect that the presence of a national plan will allow for a rapid response to a pandemic or severe biological attack? The truth is that, while the national plan is by all accounts a good one, implementing it will not be easy and it does not guarantee that every citizen will be adequately treated.

A possible reason for raising unrealistic public expectations may be related to the disaster myths discussed in Chapter 2. The simple fact is that government leaders consistently underestimate the capacity of the public to deal with the truth. There is a subtle patronization on the part of officials who display an attitude that "we know best" and a resulting decision to "protect" the public by withholding relevant facts. No one wants to be the first senior elected official to stand up before the public and say, "Our emergency preparedness **s**eriously **u**nder-**u**tilizes **c**apabilities, **k**nowledge, and **s**kills."*

Nevertheless, these unrealistic expectations can come back to haunt officials following a crisis. News reporters often joke that there are three phases to every story:

1. *What happened?* During this phase, reporters are after basic facts: What was the event? Where did it occur? Who was affected? What are you doing about it? How many were injured? These are basic, predictable questions that should be part of the jurisdiction's crisis communications plan and handled routinely by the JIC.
2. *How did it happen?* This is the "second day story" where the basic information is replaced by expert interviews and maps and diagrams showing how the incident occurred. Again, these types of questions are predictable and can be handled as routine information requests by the JIC. The JIC can also arrange for subject matter experts from the jurisdiction to provide background information as needed.
3. *Who's to blame?* The final phase of any story is an attempt to assign blame for failures that occurred. Was the incident preventable? Was the response inadequate? These types of questions may initially be asked of the JIC but will rapidly escalate to the senior elected official. Waiting until this phase begins to craft a strategic message is an almost sure guarantee of failure. Like the other two phases, this phase is predictable and can be planned for as part of a crisis communications plan.

This attempt to fix blame is actually fairly characteristic of the recovery phase. During immediate response, activities are focused on survival and meeting basic needs such as food, water, and shelter. As the crisis passes, underlying social issues are exacerbated and dissatisfaction with response failures, real or perceived, begins to surface. Witness the repeated charges of racism leveled by the citizens of New Orleans both during the response and as part of the tension between planners seeking to implement mitigation strategies and communities seeking to rebuild. These charges are in part a reflection of significant racial tension within the community prior to Hurricane Katrina.

This suggests that a major strategic task for the crisis management team is consideration of the message that it wants to send to the public. This is done, not so much to deflect blame or to shield officials, but rather to create a unifying message that allows the community to effectively recover from crisis. It has been said that, at the time of crisis, the only message people want to hear is, "It's going to be alright." This confidence must somehow be conveyed to the public—"Government is working on this; there have been problems but we are working to overcome them."

Symbolic gestures become important to the message. There is generally a defining moment in each crisis that determines public perception of the response. Officials need to recognize this moment as what it is—an opportunity to communicate their message—and take concrete action that symbolically reinforces that message. This is the reason that presidents tour disaster sites and why on-the-ground meetings with victims are more effective than fly-overs. Mayor Giuliani was highly praised for his leadership during September 11, not so much because of his extensive behind-the-scenes work, but because he was highly visible. While his undoubted leadership was critical to the success of the response, it was the public perception of that leadership that calmed and reassured citizens.

This reinforces the previous discussion that the role of senior officials is dealing with the crisis, not the tactical response. Senior officials must be seen and must be seen in a manner that emphasizes their commitment to the public. Had Michael Brown been able to articulate a unifying message in conjunction with Governor Blanco of Louisiana and Mayor Nagin of New Orleans (e.g. "We are responding with a tremendous amount of resources; there are problems, but together we are fixing them") and coupled that message with an overwhelming response at the Superdome, the symbolic pressure point, the public impression of the response to Katrina would have been decidedly different.

STRATEGIC RECOVERY ISSUES

If one accepts that a major focus of strategic response is economic restoration of the community, then it follows that a major concern of senior officials should be the transition to recovery and long-term reconstruction. This focus allows the crisis management team to guide tactical response toward this desired end state. For example, restoration of critical infrastructure will be coordinated at the tactical level, but the crisis management team establishes the priority of restoration.

An important first decision is the establishment of a governance structure for recovery. Ideally, this will already have been done through the development of a long-range recovery plan. The governance plan should identify the designated lead agency (an essential step to isolating the crisis) and the membership of the recovery task force. It should also provide for the empowerment of the task force to accomplish reconstruction goals and delineate the relationship of the task force to ongoing operational response activities. The governance structure should also provide for public oversight and involvement in the process through work groups and public hearings.

Certain policy issues are predictable by the nature of disasters. There will be the inevitable conflict between the demand to rebuild in a manner that replicates pre-disaster conditions and the desire to improve quality of life through mitigation. At the time of this writing, these very issues are playing out in New Orleans where the official plans to relocate neighborhoods to less hazardous areas are being resisted by the desire of communities to replicate their pre-disaster neighborhoods. The situation in New Orleans is a classic one: the significantly reduced population offers an opportunity for relocation and long-range hazard reduction. However, like the Burnham Plan discussed in a previous case study, the plan for New Orleans enjoys little community support and is unlikely to succeed.

Similar issues will arise as the lessons of the disaster produce recommendations for improved building codes and standards. There is usually a resistance to the extra cost and delays resulting from implementing new codes and standards. Despite its lip service to mitigation, the federal government's public assistance program has not helped the situation—it will normally not fund rebuilding to codes and standards that were not in existence at the time of the disaster. To make this happen requires a significant application of political will, something that can only be done at the strategic level.

Tactical issues can also have long-term consequences if not guided by strategy. Debris from Hurricane Iniki in 1992 forced the closure of the landfill on the island of Kauai, a facility that had been expected to be in use for another 20 years. The offshore dumping of debris from the Loma Prieta earthquake provoked substantial public outcry over concerns of environmental damage. Similar issues are emerging in New Orleans in the aftermath of Hurricanes Katrina and Rita, as demonstrated in the sidebar case study. Environmental issues, historic preservation issues, insurance issues—all of these will appear and need to be dealt with by the crisis management team. Many of these public policy issues are predictable and could be addressed through the strategic planning process before disasters occur.

Another critical area is the need for emergency legislation. It is not uncommon following a disaster to find that agencies do not have the authorities or jurisdiction necessary to adequately deal with new disaster tasks. There is usually a need to craft new ordinances to meet these needs. There may also be the need, as noted above, to quickly assess lessons from the disaster and to issue new codes and standards to assist mitigation efforts during reconstruction.

Overshadowing all of these issues, however, is the need to craft a strategy to restore economic vitality to the community as quickly as possible. Following the attacks on September 11, New York Mayor Giuliani made the restoration of the stock exchange a high priority. His staff was also immediately involved in exploring options to support businesses that had been affected, such as tax incentives, bridge loans, and federal grants. Across the country, San Francisco Mayor Willie L. Brown, Jr. and his staff created an "Open for Business" campaign aimed at encouraging the public to patronize local shops and restaurants that had seen a significant drop in business following the attacks.

One of the most critical components of economic recovery is the issue of interim and long-term housing. The sheltering of populations following an event must be seen as a continuum that stretches from immediate sheltering to some measure of permanent housing.

CASE STUDY: VILLAGE DE L'EST, LA—POOR STRATEGIC PLANNING LEADS TO CHARGES OF ENVIRONMENTAL RACISM

In April 2006, the mayor of New Orleans, the Louisiana Department of Environmental Quality (LDEQ), and the U.S. Corps of Engineers used emergency authorities to create an emergency landfill to dispose of debris from Hurricanes Katrina and Rita. In its defense against a restraining order brought by community groups, LEDQ cited the need to speed recovery in and around Orleans Parish and asked for "regulatory flexibility and consideration of the timeframe" in constructing the landfill. The result was the creation of the Chef Menteur Landfill, located a mile from the community of Village de l'Est.

The landfill is exempt from public safety protections required at other landfills, such as synthetic liners, groundwater monitoring systems, and leachate collection systems. The landfill consists of a 35-foot deep pit with a natural clay bottom 10 to 15 feet deep in what is essentially a wetlands area where the water table is one to four feet below the surface. An 80-foot canal adjacent to the landfill is used by many residents to water lawns and gardens and separates the landfill from the Bayou Sauvage National Wildlife Refuge, the largest urban wildlife refuge in the country. The area is considered the most susceptible part of the city to storm surge and was the last area to be de-watered following Hurricane Katrina. Debris trucked to the landfill passes through Village de l'Est at the rate of 1,000 truckloads a day, creating concerns over debris from the uncovered trucks being scattered throughout the community.

In theory, the Chef Menteur Landfill does not allow the dumping of domestic and commercial appliances, bacteria encouraging materials or hazardous, liquid, infectious, commercial, industrial, or residential wastes. In actual practice, the Environmental Protection Agency believes that only 20 to 30 percent of such materials can actually be filtered out of the debris. Further, inspectors were recently removed from the site and LEDQ and the site operator will not allow testing of the site by outside experts.

While most of eastern New Orleans is African-American, Village de l'Est has a large community of Vietnamese-Americans who claim their concerns were ignored during the 30-day public comment period when the landfill was proposed. They have formed a community action group, Citizens for a Strong New Orleans East (CSNOE), and have allied with the Louisiana Environmental Action Network to oppose the landfill on the basis of its potential impact on public health. CSNOE believes that there is more than adequate capacity in five existing landfills and that a better organized system of collection using staging areas and night hauling would be both environmentally sound and faster than the cur-

(continued on next page)

(continued from previous page)
rent system. The group claims that Mayor Nagin lifted zoning rules that had defeated two previous attempts to create landfills near Village de l'Est because the landfill operator agreed to donate 22 percent of the site's revenue to the city. They consider it significant that the mayor halted dumping while the site was tested for the week prior to the mayoral election in May and promptly resumed dumping after the election. CSNOE also suggests that the operator is knowingly covering up problems by refusing to allow outside testing and by refusing to allow expert observers full access to the site.

The issue of debris clearance is one that occurs following any major disaster. The Chef Menteur Landfill is an example of the inherent conflict between the need to protect public health and the environment and the desire to rebuild quickly. It is not an isolated case. Similar community resistance to debris removal can also be found in Waggaman, a predominantly African-American community that houses the River Birch Landfill and a construction and demolition debris site where a considerable amount of debris from Hurricane Katrina has been dumped. Heavy traffic by uncovered trucks has caused roads to deteriorate and littered the community with trash, raising concerns over asbestos and other health risks.

Debris disposal is a response generated need. It can therefore be anticipated and appropriate strategies and plans developed to meet anticipated requirements. Not only is there no evidence that such was the case in New Orleans, there was clearly a failure of strategic crisis management in creating the Chef Menteur Landfill. The issue is not whether the landfill is needed or whether all possible alternatives were considered. Rather, the issue is the failure to anticipate strong community outrage and to have a strategy in place to alleviate concerns by involving the community in the decision-making process. Community outrage over a landfill is clearly to be expected under most circumstances; when tensions are running high following a crisis, it is inevitable, particularly if the landfill is placed in economically disadvantaged or minority neighborhoods and perceived to lack appropriate public health safeguards. Further exacerbating this perception is historical experience: debris from Hurricane Betsy in 1965 was deposited and burned atop a former landfill that later had to be declared a "Superfund" toxic cleanup site.

Natural community resistance to landfills, environmental concerns, and historical experience made the issue of debris removal following Hurricanes Katrina and Rita a significant policy issue. If this issue had been identified and addressed in the early days following Hurricane Katrina, there would have been less need to invoke emergency authorities and confront community outrage almost nine months later.

Unfortunately, most jurisdictions focus solely on the tactical issue of sheltering without accepting that congregate shelters address only a fraction of those who are displaced. Many victims lodge with family or friends, and a small segment of the displaced have resources through insurance. However, this wide dispersion of population can have a severe impact on economic recovery, as is being demonstrated in New Orleans.

An initial strategic concern is to bring citizens back to the affected area through the provision of temporary housing or quick repairs to damaged homes. However, interim or temporary housing has its own set of problems. Most jurisdictions do not have excess housing stock and have little capacity to absorb displaced populations. This suggests that victims must either be sheltered in other communities, as was done after Katrina, or that some form of temporary housing must be constructed similar to those in Kobe or in San Francisco in 1906. However, the modern version of using mobile homes may actually add to overall community risk and have unintended social impacts. Further, temporary facilities have a way of becoming semi-permanent in the absence of an aggressive strategy to close them as quickly as possible. This, in turn, puts pressure on the jurisdiction to expedite reconstruction at the expense of mitigation.

Economic recovery is a critical component of strategic response. There is an unfounded expectation on the part of many jurisdictions that the federal government will provide major assistance in rebuilding a community. Unfortunately, federal recovery programs are extremely limited. In the case of individual assistance, they consist primarily of low interest loans provided through the Small Business Administration. Public assistance is provided through a series of programs that require the jurisdiction to expend funds that are then reimbursed. In practice, obtaining reimbursement comes years after the disaster and may represent only a fraction of the jurisdiction's actual cost. These programs usually have a 25 percent cost share and federal staff work hard to reduce the amount they must reimburse. A potential source of revenue is insurance, but this is usually only pertinent to the private sector, as most jurisdictions tend to be self-insured.

Significant federal assistance is not provided through traditional reimbursement programs. Instead, it is provided through special legislation and through hardheaded negotiation with agencies such as FEMA. The issue is a political one and, therefore, in the province of strategic response. Part of the planning for recovery should the formation of a recovery team to deal with these issues separate from the routine reimbursement programs.

CATASTROPHIC EVENTS

For the most part, this book has not distinguished between planning for disasters and planning for catastrophic events. This is because, for the most part, the processes are similar and, to the tactical responder, there is little difference between them. A citizen affected by an event doesn't care whether it is an emergency, a disaster, or a catastrophe.

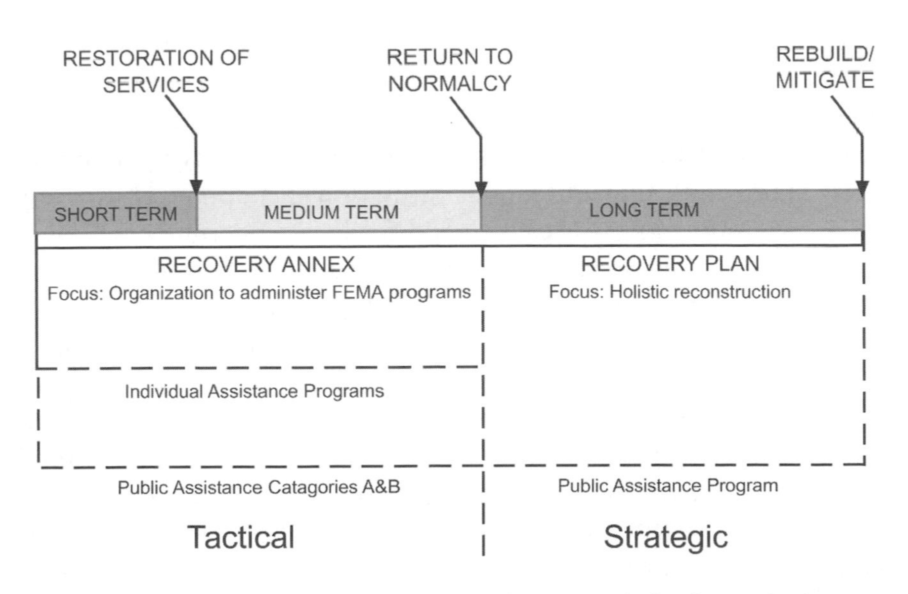

Figure 10.3 Recovery timeline distinguishing between tactical and strategic elements.

Similarly, responders don't really consider the difference between catastrophic events and disasters. When an event occurs, all necessary resources are applied—nothing is held back on the basis that the event is "merely" a disaster. This supports the contention that catastrophic planning is strategic, rather than tactical in nature.

In developing plans for catastrophic events, the focus is not solely on the jurisdiction's response, but on regional strategies. Mutual aid can no longer be drawn from adjacent jurisdictions but must be mobilized on a statewide level. Evacuees must be moved further than normally called for in tactical disaster plans. Coordination, rather than command and control, becomes even more critical than in disasters, particularly as federal agencies become more involved in providing resources.

This suggests two things of concern to the crisis management team. The first is that the jurisdiction must be adequately represented in regional decision making. This goes beyond sending liaisons to the county or state EOC to coordinate tactical issues. Instead, it suggests the need for strong political representation in dealing with state and federal representatives as decisions are made regarding funding and availability of programs.

This is closely tied to the second concern: the need to assess long range needs and the potential impacts of state and federal programs. In considering the Lisbon earthquake of 1755 discussed in Chapter 1, one is struck by the single-mindedness of the Marquis of Pombal in preserving his workforce by limiting evacuation. If one compares this to the scattering of the evacuees of Hurricane Katrina over the entire United States, and the subsequent reduction in population in New Orleans, one immediately sees the need for local government involvement in deciding how to handle evacuees. This, in fact, seems to be the case in Florida where the state is creating host communities as part if its catastrophic planning to keep evacuees as close to home as possible.

Another example of long-range impacts on population is the use of trailers as temporary housing. While a quick and efficient way of meeting immediate needs, trailers are extremely vulnerable to natural disasters. The use of trailers in hurricane-prone areas actually creates increased risk in subsequent disasters. The decision to use trailers is usually made unilaterally by the federal government in response to public demand. However, since their use creates long-range problems for the community, it therefore constitutes a significant strategic issue.

The importance of local government engagement in making strategic decisions cannot be overstated. Without the forceful involvement of local government and strong political allies in state and federal legislatures, state and federal agencies have a tendency to implement "business as usual" practices and will mirror strategies that worked in other events. Cultural sensitivities are not always taken into account. An example of how active involvement in the decision-making process can benefit a local community is shown in the sidebar case study on FEMA's activities in Yap State in 1991.

This brief discussion suggests two areas of strategic planning for the jurisdiction. The first is the crafting of an overall regional strategy for dealing with the impact of catastrophic events. Such a strategy must take into account the jurisdiction's role as a provider of services as well as that of a user. In a catastrophic event, seemingly remote jurisdictions may find themselves supporting evacuation corridors with traffic control, refueling points, or rest areas and may be tasked to provide host sheltering either through congregate facilities or local housing stock. This suggests that the current system for declaring disasters and seeking reimbursements for disaster related costs might need to be revised. Reimbursement of host jurisdictions posed a significant problem during the Katrina evacuation.

The second strategic planning issue is the development of a strategy to marshal legislative support and ensure adequate representation in the decision-making process. Many jurisdictions already have components of this system in place such as political lobbyists and public information officers. However, in a time of crisis, the actions of these groups must be coordinated under a cohesive strategy. While some of the issues arising may be agent generated, for the most part, they will be response generated and can therefore be anticipated.

CONCLUSION

Previous chapters have made the case for viewing response from three levels: operational, tactical, and strategic. Most emergency management programs address the first two but only marginally consider the third level. Yet it is the strategic level that ultimately determines the long-range success of the response and guides the community back to

CASE STUDY: SUPER TYPHOON OWEN IN YAP STATE: JOINT DECISION MAKING PRESERVES CULTURAL IDEALS

In late November 1990, Super Typhoon Owen struck the Federated States of Micronesia, a U.S. protectorate in the northwestern Pacific. With winds exceeding 150 mph, the storm destroyed as much as 95 to 99 percent of the housing and 80 to 90 percent of the crops on some of the islands. Hardest hit were the states of Yap and Chuuk. Under the terms of the protectorate, the Federal Emergency Management Agency coordinated the response.

FEMA's initial concern was to provide individual assistance by providing a grant to each of the victims. Chuuk State willingly accepted the funding but Governor Petrus Tun of Yap State expressed concerns that the money would be misused and create social problems in a state trying to preserve its cultural heritage. Yap State is proud of its cultural heritage and is committed to preserving its traditional ways. Governor Tun was concerned that a large influx of outside money would disrupt traditional community values. A study of Chuuk State some months later, by a group of college students from Guam, suggested that relief funds there had been used to purchase luxury items and for travel to Guam rather than for necessary supplies, justifying the governor's concern.

However, the governor's refusal to accept funds created a problem for FEMA. U.S. law demanded that comparable assistance be provided in each state. Fortunately, the FEMA federal coordinating officer, A. Roy Kite, was highly experienced and culturally sensitive. There was a mutual respect between Kite and Governor Tun and a compromise was soon reached. Instead of the funding that would jeopardize the traditional culture, FEMA agreed to purchase and distribute family kits to replace items lost in the typhoon. The governor's staff worked with FEMA to develop lists for the kits that included items such as cooking pots, fish hooks, lanterns, axes, and other household items. The kits met the need for assistance without the influx of cash that might have led to significant social changes.

FEMA's next concern was the development of permanent housing. FEMA had developed plans for typhoon-resistant housing that was culturally acceptable to Pacific Islanders and planned to provide the materials and funding for labor necessary to construct the housing. Again, Chuuk State was agreeable, but Governor Tun objected to paying his people to construct housing. The tradition in Yap was that the community helped rebuild a house that was lost through fire or other catastrophe, and the governor was concerned that paying citizens to rebuild their neighbor's home would signal the end of this tradition. Instead, he asked for technical assistance in building the typhoon resistant houses.

(continued on next page)

(continued from previous page)

Again, Governor Tun and Kite were able to work out an acceptable compromise. FEMA provided building materials and tools and deployed a team of SEABEES to travel to the various island groups to train local builders. The local teams then used the tools and materials to build the houses needed by their community. The result was a quick replenishment of the housing stock with typhoon-resistant structures while preserving the cultural traditions of Yap State. Governor Tun was successful both in reconstructing his severely damaged state and in preserving Yapese cultural traditions. He accomplished this by understanding that the crisis was not the funding of reconstruction but the preservation of cultural traditions while expediting reconstruction. By involving himself in strategic decision making, he was able to modify the "business as usual" approach by FEMA and craft a recovery strategy that was effective and mutually agreeable.

an acceptable end-state. While best practices documents such as NFP1600 and State and Local Guide 101 address the tactical, there is little emergency management guidance for the strategic. There is, however, a considerable body of literature on the management of crisis and these principles can be directly translated to the disaster arena.

The proper role of senior managers is management of the crisis, not tactical response. The real issues affecting the community are related to the economic long-term needs generated by the disaster and only senior officials have the broad social and political perspective necessary to identify and manage these issues. Crisis management, the identification and resolution of long-range strategic issues, is the principal role of senior officials in a disaster.

**With thanks to Dennis Dura, New Jersey Department of Human Services.*

CONCLUSION

Emergency management is not what you do; it is who you are.
—Henry Renteria
Director, California Governor's Office of Emergency Services

The preceding chapters have attempted to make a case for a new way of approaching emergency management. Traditional emergency management has its roots in the early civil defense programs of the cold war era and is still heavily dominated by the military influence of those times. This has resulted in many programs that are plan-centric and focused primarily on tactical issues. It has also led to the institutionalization of many disaster myths that have been continually challenged by social science. These plans assume that a breakdown of social norms is inevitable in a disaster and that only strong, centralized command and control can overcome the resulting chaos.

These plans create an Achilles heel for many programs. In *Managing the Unexpected*, researchers Karl Weick and Kathleen Sutcliffe suggest that plans may create a false sense of security by lulling planners into a sense of predetermination, creating a disregard for the dangers of the unexpected. Lee Clarke in *Mission Improbable* and Erik Auf der Heide in *Disaster Response* raise similar cautions. The focus on the development of a plan rather than on community resilience is the major reason many emergency management programs fail to build an adequate capacity to respond.

Rather than a plan-centric approach, this book suggests that emergency plans are a component of a holistic program that provides benefit to the community. The goal of emergency management programs is not the management of response but the management of risk, a much broader context. Emergency management programs provide a mechanism for assessing risk and developing and implementing strategies aimed at reducing or eliminating risk and for building the capacity to protect the community from the unexpected.

It is in this area of strategy development that many programs fail. Emergency managers have been doing rote work for years, allowing programs to be driven by grant requirements. Plans are written to conform to grant requirements. Exercises are conducted to conform to grant requirements. Even the size and composition of many emergency management offices have has been determined by grant requirements. Somehow along the way the idea of developing risk-based strategies was lost and these grant activities became an end in themselves.

Despite the lip service given to the four-phase comprehensive emergency management model, emergency managers have narrowed their focus and directed their work almost solely toward response and almost exclusively to the development of the emergency operations plan. Strategic issues related to mitigation and recovery are barely addressed. Continuity planning is completely foreign to most emergency managers.

This dynamic can change if emergency managers are viewed not as technocrats but as program managers with the responsibility for crafting strategies for community resilience. When viewed from this perspective, the emergency manager's position changes from one that is viewed as a necessary evil and expense to one that is a key player in building community resilience and assisting in the management of risk. This further suggests that the emergency manager should not write emergency plans but should ensure that emergency plans are written. The difference is subtle but significant—emergency managers are program managers who specialize in managing part of the community's risk rather than technicians who write specialized plans. A manager prepares the community as a whole by engaging all the resources of the community under a shared vision of resilience.

The capacity to respond to crisis incorporates more than the traditional operational response functions found in emergency management texts. This capacity is dependent on a collaborative process that can be expanded to include emergent organizations and that has the ability to deal with new disaster tasks. It places a premium on improvisation and creative problem solving. It also must recognize that roles differ at various levels of response and that successful response must make use of the expert skills of the responders.

This suggests that plans need to be considered in relationship to each other on the basis of time of implementation. Instead of a response plan that considers life safety and another plan for continuity and a third for recovery, it may make more sense to have a response plan that incorporates elements of safety, continuity and response. Breaking down the stovepipes that separate plans can eliminate confusion and competition for resources during crisis. This also suggests that one must distinguish between the need for command and control functions at the field level of response and the need for coordination at the tactical level.

Ultimately, the emergency management program must be perceived as adding value to the community. This means that it must be consistent with community values and vision of the future. This suggests that emergency management plans should be community plans that include all mechanisms by which the community responds to crisis. This concept of value-added implies a change in the role of emergency managers. Instead of being experts in tactical and operational response, emergency managers must assume the role of program coordinators. This means that they are expected to assemble appropriate groups of stakeholders and facilitate the development of strategies for response, recovery and mitigation. It means accepting responsibility for overseeing the strategic plans that implement this vision. It means providing leadership, not just technical expertise.

This leadership role must be built on a solid knowledge of social science research and historical disasters. It also requires the technical skills related to traditional emergency management concepts and methods and the latest national guidance. It requires an understanding of group dynamics and interactions, of meeting facilitation skills, of writing and speaking skills. But most of all, it requires passion—a passion that allows for tilting at windmills and challenging conventional wisdom, that draws others to the table and makes them commit to a shared vision of community resilience.

BIBLIOGRAPHY

Answering the Call: Communications lessons learned from the Pentagon attack, Public Safety Wireless Network Program, Washington D.C.

Auf der Heide, Erik, 1989. *Disaster Response: Principles of preparation and coordination*, C.V. Mosby Company, St. Louis, MO.

Ball, Philip. Mass graves not necessary for tsunami victims: Rapid burial to avert health risks a myth. *Nature*, 2005.

Barsky, Lauren, Joseph Trainor and Manuel Torres, 2006. *Disaster Realities in the Aftermath of Hurricane Katrina: revisiting the looting myth*, Quick Response Report 184, University of Colorado Natural Hazards Center.

Burns, Tom and G.M. Stalker 1961. *The Management of Innovation*, Tavistock, London.

Characteristics of Effective Emergency Management Organizational Structures, Public Entity Risk Institute, Fairfax, VA.

Clarke, Lee, 1999. *Mission Improbable: Using fantasy documents to tame disaster*, University of Chicago Press, Chicago, IL.

———. Worst case thinking: an idea whose time has come. *Natural Hazards Observer* Volume XXXIX Number 3, January 2005, University of Colorado, Boulder, CO.

DiMartino, Christina. Picking up the pieces part ii, *Waste Age*, December 1, 1999.

Drabek, Thomas E., and Hoetmer, Gerard J., (eds), 1991. *Emergency Management: Principles and practices for local government*, International City Management Association, Washington, D.C.

Dynes, Russell R. 2003, The Lisbon earthquake in 1755: the first modern disaster, Preliminary Paper #333, Disaster Research Center, University of Delaware.

———, 1995. The impact of disaster on the public and their expectations, Preliminary Paper #234, Disaster Research Center, University of Delaware.

———, Noah and Disaster Planning: The cultural significance of the flood story, University of Delaware.

———, 1994. Community Emergency Planning: False assumptions and inappropriate analogies, *International Journal of Mass Emergencies and Disasters*, August 1994, Vol. 12, No. 2, pp.141-158.

Eberwine, Donna. Disaster myths that just won't die, *Perspectives in Health*, Volume 10, Number 1, 2005.

Ethridge, Frank. Don't trash our neighborhood: A New Orleans East Vietnamese community battles a landfill in its midst, *Best of New Orleans*, June 20, 2006.

Fink, Steven, 1986. *Crisis Management: Planning for the inevitable*, American Management Association, New York, NY.

Grunwald, Michael and Glasser, Susan B. Brown's turf wars sapped FEMA's strength, *Washington Post*, December 23, 2005.

Haas, J. Eugene, Robert W. Kates, and Martyn J. Bowden (eds). 1977. *Reconstruction Following Disaster*. Cambridge Massachusetts: MIT Press.

Hiller, Andrew, 2000. *Business Continuity: Best Practices*, Rothstein Associates, Inc., Brookfield, CN.

Holistic Disaster Recovery: Ideas for Building Local Sustainability After a Natural Disaster, 2001, Public Entity Risk Institute, Fairfax, VA.

Jablonowski, Mark. Do catastrophe models mislead? *Risk Management*, Risk and Insurance Management Society, Inc.

Kaplan, Tamara. The Tylenol crisis: how effective public relations saved Johnson & Johnson, Pennsylvania State University.

Kelly, Jack. No shame: The federal response to Katrina was not as portrayed, *New Orleans Times Picayune*, September 11, 2005.

Kelly, John, 2005. *The Great Mortality: An intimate history of the Black Death, the most devastating plague of all time*, Harper Collins, New York, NY.

Lewis, Ralph G. 1988. Management Issues in Emergency Response, *Managing Disaster*, Louise Comfort, editor, Duke University Press, Durham, MA.

Lindell, M.K. and R.W. Perry and C.S. Prater 2005. Organizing response to disasters with the incident command system/incident management system (ICS/IMS), International Workshop on Emergency Response and Rescue, October 31-November 1, 2005.

Management of Dead Bodies in Disaster Situations, Pan American Health Organization, Washington D.C., 2004.

Minutaglio, Bill 2003. *City on Fire: The forgotten disaster that devastated a town and ignited a landmark legal battle*, Harper Collins, New York, NY.

Monday, Jacquelyn. Building back better: creating a sustainable community after disaster, *Natural Hazard Informer*, January 2002, University of Colorado, Boulder, CO.

Monmonier, Mark, 1997. *Cartographies of Danger: mapping hazards in America*, University of Chicago Press, Chicago, IL.

Moyer, Heather. Landfill angers East New Orleans, *Disaster News Network*, June 21, 2006.

Phillips, Brenda C., 2003. Disasters by discipline: necessary dialogue for emergency management education, Jacksonville State University, Jacksonville, FL.

Petterson, Jeanine, 1999. A Review of the Literature Programs on Local Recovery from Disaster (Working Paper #102) Public Entity Risk Institute, Fairfax, VA.

Platt, Rutherford H., 1999. *Disasters and Democracy: The politics of extreme natural events*, Island Press, Washington D.C.

Quarantelli, E.L., 2005. The earliest interest in disasters and the earliest social science studies of disasters: a sociology of knowledge approach (draft), Disaster Research Center, University of Delaware.

———, 2005. Catastrophes are different from disasters: some implications for crisis planning, Social Science Research Council, New York, NY.

———, 2003. A half century of social science disaster research: selected major findings and their applicability, Preliminary Paper #336, Disaster Research Center, University of Delaware.

———,2000. Disaster planning, emergency management and civil protection: the historical development of organized efforts to plan for and to respond to disasters, Preliminary Paper #301, Disaster Research Center, University of Delaware.

———,1997. Research Based Criteria for Evaluating Disaster Planning and Managing, Disaster Research Center, University of Delaware.

———,1989. How individuals and groups react during disasters: planning and managing implications for EMS delivery, Preliminary Paper #138, Disaster Research Center, University of Delaware.

———,1979. The consequences of disasters for mental health: conflicting views, Preliminary Paper # 62 Disaster Research Center, University of Delaware.

Rhoads, Christopher. Cut off: At center of crisis, city officials faced struggle to keep in touch, *Wall Street Journal*, September 2005.

Rudman, Warren B., Chair, 2003. *Emergency Responders: Drastically underfunded, dangerously unprepared*, Council on Foreign Relations, New York, NY.

Ryan, William and Robert Pitman, 2000. *Noah's Flood: The new scientific discoveries about the event that changed history*, Touchstone, New York, NY.

Sandman, Peter M., 1993. *Responding to Community Outrage: Strategies for effective risk communication*, American Industrial Hygiene Association, Fairfax, VA.

Schneider, Robert. A strategic overview of the "new" emergency management, University of North Carolina, Pembroke, NC.

Schoch-Spana, Monica, 2005. Public responses to extreme events—top 5 disaster myths, University of Pittsburgh Medical Center, Pittsburgh, PA.

Schwab, Jim, 1998. *Planning for Post-disaster Recovery and Reconstruction*, American Planning Association, Chicago, IL.

Shea, Christopher. Up for Grabs: Sociologists question how much looting and mayhem really took place in New Orleans, *The Boston Globe*, September 11, 2005.

Steinberg, Ted, 2000. Acts of God: *The unnatural history of natural disaster in America*, Oxford University Press, New York, NY.

Sylves, Richard. A précis on political theory and emergency management, University of Delaware, Newark, DE.

Thevenot, Brian and Gordon Russell. Rumors of deaths greatly exaggerated, *New Orleans Times Picayune*, September 26, 2005.

Tierney, Kathleen J., 2005. Recent developments in U.S. Homeland Security policies and their implications for the management of extreme events, University of Colorado, Boulder, CO.

———,1993. Socio-economic aspects of hazard mitigation, Disaster Research Center, University of Delaware.

———, 1993, Disaster preparedness and response: research findings and guidance from the social science literature, Disaster Research Center, University of Delaware.

Tierney, Kathleen J.; Michael K. Lindell and Ronald W. Perry, 2001. *Facing the Unexpected: Disaster preparedness and response in the United States*, Joseph Henry Press, Washington D.C.

Tworney, Steve, Carol D. Leonnig and Petula Dvorak. District unprepared to cope with attack: police improvised, no broadcast made. The Washington Post, September 17, 2001.

Tufte, Edward R., 1997. *Visual Explanations: Images and quantities, evidence and narrative*, Graphics Press, Cheshire, CN.

Unseating the myths surrounding the management of cadavers, *Disasters, Preparedness and Mitigation in the Americas*, Issue 93, October 2003.

Ward, Kaari, Editor, 1989. *Great Disasters: Dramatic true stories of nature's awesome powers*, Readers Digest, Pleasantville, NY.

Weick, Karl E. and Sutcliffe, Kathleen M., 2001. *Managing the Unexpected: assuring high performance in an age of complexity*, Jossey-Bass, San Francisco, CA.

Woodbury, Glen. Measuring prevention, *Homeland Security Affairs*, Volume 1, Issue 1, Article 7, Summer 2005.

INDEX